C0-CDK-387

SOIL GEOMORPHOLOGY

SOIL GEOMORPHOLOGY

Raymond B. Daniels
North Carolina State University
Raleigh, North Carolina

Richard D. Hammer
University of Missouri
Columbia, Missouri

JOHN WILEY & SONS, INC.

New York • Chichester • Brisbane • Toronto • Singapore

Library of Congress Cataloging-in-Publication Data:

Daniels, Raymond Bryant, 1925–
Soil geomorphology / Raymond B. Daniels and Richard D. Hammer.
p. cm.
Includes bibliographical references.
ISBN 0-471-51153-6 (cloth)
1. Geomorphology. 2. Soils. I. Hammer, Richard D. II. Title.
GB406.D26 1992 92-3789
551.4′1—dc20 CIP

Printed and bound in the United States of America by Braun-Brumfield, Inc.

10 9 8 7 6 5 4 3 2 1

To our wives
Irene and Jennifer

ACKNOWLEDGEMENTS

This textbook is about geomorphology and soils in their natural settings. It is appropriate that I recognize in it those persons who have most influenced my approach to soils in the field. It is no coincidence that all of them were "field men."

When I first began in this profession, Dr. Bob Gilmore stressed the importance of a quantitative approach to field work. The late Burt Ray struck in me the first sparks of appreciation for the dynamic nature of soils. The late Bud Giese taught me to trust my judgment and instincts in the field. Dr. George Buntley stressed to me the relationship between soil water and soil morphology. Dr. Glen Smalley enhanced my awareness of landforms as determinants of forest composition, and Dr. Larry Parks gave me the opportunity to be a research scientist. Some part of each of them is in this book. I hope they share its perspectives.

R.D. HAMMER

I want to thank Dr. R.V. Ruhe for his many hours of patient explanation and instruction. He made the field of geomorphology and soils an interesting and exciting area to investigate. I also wish to thank Dr. E.E. Gamble for several years of intensive, interesting field work. The arguments were stimulating and fruitful.

R.B. DANIELS

We want to thank Dr. Jim Richardson, the late Dr. Walter Wheeler, and Dr. Egon Klampt and Mr. Ralph Heath for their critical review on the manuscript and many helpful suggestions. We also thank Dr. G. J. Buntley for his excellent work in creating our cover illustration.

RBD and RDH

CONTENTS

Preface xvii

1 Introduction 1

Field Investigations 2
- Stratigraphy 3
- Geomorphology 3
  - Erosional Surfaces 5
- Hydrology 7
- The Soil System 8

References 9

2 Stratigraphy 10

Introduction 10
- Soil Variability 10
  - Data Analysis 11

Stratigraphic Principles and Concepts 13
- Definitions 13
- Superposition 14
- Relative Dating 15
  - Terraces 17
  - Marine and Glacial Sediments 18
  - Mud Flows 21
- Geochemical Dating Methods 21

References and Bibliography 23

3 Textural Characteristics of Soil Materials 25

Introduction 25

Terrestrial Environments 26
- Glacial Drift 27
  - Depositional Environments 28
  - Environments and Landforms 30
- Eolian Deposits 34
  - Eolian Sands 35

Textural Properties 35
Distribution and Form 36
Loess 39
Texture 39
Other Properties 41
Local Thinning 42
Soils 42
Parna 43
Aerosols 44
Additional Reading 44
References 45

4 Fluvial Systems **48**

Introduction 48
Braided Stream Deposits 48
Meandering Stream Deposits 50
Environments 50
Application 56
Map Unit Variability 57
Interpretations for Soils 58
Fans and Associated Deposits 58
Fans 59
Environments 60
Playas 65
Prediction Value 65
Lakes 67
Shore and Shoreline Features 68
Freshwater Lakes Dominated by Sediment Input 68
Saline Lakes 68
Sediment Starved Lakes 69
Lacustrine Sediments and Soils 70
Additional Reading 73
References 73

5 Hillslope Sediments **76**

Introduction 76
Distribution and Properties 78
Distribution 78
Properties 79
References 84

6 Transitional Environments and Terrigenous Marine Shelf Sediments **86**

Introduction 86
Deltas 86
River-Dominated Deltas 88
Wave- and Tide-Dominated Deltas 88
Estuaries 90
Barriers and Associated Environments 93
Lagoon Deposits 94
Barrier Deposits 96
Transitional Systems and Soils 101
Shoreface and Terrigenous Marine Shelf Sediments 104
Beach and Shoreface 107
Shelves 108
Facies 108
Transgressive-Regressive Facies 109
Summary 110
Additional Reading 111
References 112

7 Volcanic Materials **115**

Introduction 115
Kinds of Volcanic Rocks 115
Fragmented Volcanic Rocks 116
Terminology 118
Pyroclastic Deposits 119
Pyroclastic Flow Deposits 119
Pyroclastic Surge Deposits 123
Pyroclastic Fall Deposits 124
Pyroclastic Materials and Soils 124
References 125

8 Saprolite **127**

Regolith Properties 127
Areal Distribution and Thickness 128
Characteristics 129
Rate of Formation 133
References 134

9 Geomorphology **136**

Introduction 136
Geomorphic Surfaces 136
Definitions 136
Types of Surfaces 137
Constructional Surface Criteria 137
Surface Properties 139
Dating Geomorphic Surfaces 139
Hillslope Nomenclature 141
References 143

10 How Landscapes Evolve **144**

Introduction 144
Theories of Landscape Evolution 144
Peneplanation 145
Parallel Slope Retreat 146
Dynamic Equilibrium 147
Backwearing 149
Nine-Element Landscape 149
Applying Landscape Models to Field Research 150
Evolution of a Theoretical Landscape 151
Application 157
References 158

11 Rates of Denudation **159**

Methods of Measurement 159
Erosion Rates 159
Caution in Interpretation 160
Late Holocene Erosion Rates 161
General Relationships 161
Small Watershed Studies 164
References 167

12 Streams **169**

Introduction 169
Degrading and Aggrading Systems 171
Examples 172
Truncated Ephemeral Channels 173
Interpretation For Soils 174
References 177

13 Hillslope Processes and Mass Movement **179**

Introduction 179
Hillslope Erosional Processes 180
Rainfall 181
Slope Factors 182
Slope Shape 182
Computer Modeling of Slope Changes 183
Problems in Quantifying Hillslope Evolution 185
Mass Wasting 185
Types 185
Soil Creep 186
Terracettes 187
Solofluction 187
Earth Flows and Mud Flows 187
Summary 189
Additional Reading 192
References 193

14 Time and Soil Formation **195**

Introduction 195
Soil Morphology as a Time Indicator 195
Time—Relative, Effective and Absolute 196
Relative Time 196
Effective Time 197
Soil Age 200
References 202

15 Hydrology **203**

Introduction 203
Basic Groundwater Hydrology 204
Flow Nets 206
Complications 210
Near-Surface Hydrology and Soils 213
Hillslope Hydrology 218
Theory 220
Other Factors 22
Application to Soils 223
Summary 227
References 227

Index **231**

PREFACE

The authors believe that one needs a knowledge of the geologic processes that shaped the landscape to understand soil distribution, properties and genesis. All landscapes are somewhat different, yet several features reappear in each. These similarities allow one to transfer ideas, usually with modification, from one area to another. The transfer of ideas between landscapes is difficult unless we use the basic relations between landscape processes and soils. There is a close relationship between geologic processes and soils. Soils are only the thin upper part of a complex system controlled in large part by stratigraphy, geomorphology and hydrology.

Our text has three major sections: stratigraphy, geomorphology and hydrology. In each of these sections we will attempt to provide the reader with the fundamentals needed for a basic understanding of the soil landscape. For each section scores of volumes are provided as references.

Few scientists, if any, are expert in all parts of their specialty. We can only hope to give the reader the ideas that are useful tools for field investigations of soils. Chapters 2, 5 and 9 are examples of ideas used by geologists that are of value to soil scientists. The remaining chapters will be illustrations of more narrow interest. For example, if one is working in delta systems, then Chapter 6 will be pertinent. If one is working in areas of active mass movement, then Chapter 13 should be useful. In each case, we have borrowed heavily from textbooks on the particular subject because they are easier for field scientists to obtain than many scientific journals. Also, the textbooks summarize a mass of data and help the reader understand general relations. We have only attempted to outline each subject, not give it an exhaustive treatment; however, we have cited much of the original literature for those interested in digging further into that particular subject.

A Ph.D. student with 10 years field experience mapping soils once commented after analyzing a watershed that he had never fully appreciated the dynamic nature of the soil system. Indeed, soils vary in space and time and are dynamic systems. Soil research often is a two dimensional or three-dimensional investigation of a multidimensional resource. Indeed, soils vary in space and time and are dynamic systems. Two-dimensional sampling ignores the spatial relationships of the resource. Three-dimensional sampling considers spatial soil properties, but ignores temporal relationships. Soil profiles are more easily understood when observed under the range of annual temperature, moisture and biological activity.

Soil morphological and chemical features are a record of past and current processes. Thus one cannot hope to interpret soil systems accurately without an understanding of how the landscape and soils have coevolved over time. Investigation of multiple profiles in the landscape helps one to appreciate current and past dynamics of the soil system.

To understand a soil system, therefore, one must investigate that system as completely as possible because the flux of materials and energy through the systems had a profound effect on individual soils. Too often, however, the authors have listened to colleagues discussing the genesis and taxonomy of a recently exposed soil profile without considering the genesis of the surrounding landscape. Soil taxonomy seems to have become a means to an end for many scientists. Discussions of the genesis and morphological features of the soil often involve only the taxonomic control section which, in many landscapes, is only the surface of the zone of biological activity, a control section much thinner than the rooting zone of perennial plants and vadose zone of water movement. Although soil taxonomy is an important tool for soil classification and a means to group soil bodies, one disadvantage of soil taxonomy is that it emphasizes the pedon at the expense of its relationship to the surrounding landscape.

The framework from which one views an object or system effects one's relationship with that object. The early soil scientists, for example, considered soils as weathered rock. This idea of soil inevitably affected soil research and interpretations. Our primary purpose in this text is to consider the soil as a component of a larger system: Stratigraphy and the past and current processes that shaped the landscape and affected material distribution in that landscape influence soils. This is but one perspective of soil systems, but is one often overlooked.

We hope this text will awaken in our colleagues a heightened awareness of the importance of landscape genesis in soil development. We can improve our ability to interpret the effect of man's activities on soils by considering the soil as the upper part of a more complex system.

1 Introduction

Soil geomorphology is the application of geologic field techniques and ideas to soil investigations. Our ideas are an expansion of those expressed by Jenny (1941) in his classic book on the factors of soil formation, although we use a different terminology and will add detail to some of his ideas. We use stratigraphy as a component of parent materials; geomorphology encompasses parent material origin, topography and time. Jenny discussed some aspects of soil hydrology, but our emphasis will be greater in scope and detail. We consider hydrology to be the primary driving force in soil formation, although stratigraphy and geomorphology control the systems' hydrology.

Soil geomorphology is of value to soil scientists because soils are good integrators of several factors in their past and present environments. An idea, or working hypothesis, of the processes responsible for soil conditions at a site is necessary for almost all field work. Without an idea to test we spend too much energy and time collecting irrelevant data. One who does not know a field site's stratigraphic and geomorphic location has little hope of understanding the active or inactive processes responsible for the soil development. Also there is little hope of accurately predicting the areal extent of similar soils. Many research soil scientists use soil map units as a basis for their work. Yet a soil map unit includes a variety of stratigraphic and geomorphic conditions, though most properties are within the ranges of one or two soil series.

Surface texture and water available during the growing season are properties that vary systematically with position on the landscape. With enough experience in an area, a field scientist can predict where the major textural changes occur and what areas will be periodically wet. Can one predict similar properties in a landscape with different sediments, geomorphology and soils? Maybe—but such predictions will be more accurate with an understanding of why these properties occur at specific landscape positions. Understanding the processes that developed a landscape and their interaction with stratigraphy and geomorphology allows transfer and application of this knowledge to a wide range of landscapes. Landscapes may look much different, but many processes controlling them are the same. Our purpose is to present some geologic fundamentals so the reader can understand why soil properties vary, and with this framework, can ask the questions that will help solve specific soil problems.

This text emphasizes the materials, geomorphology and hydrology that influence soils. We use three simple ideas throughout:

1. The law of superposition states that younger beds overlie older beds. Application of this law is fundamental to understanding the age relations and probable areal distribution of sediments or surface materials.
2. The law of ascendancy and descendency states that higher surfaces are older than lower surfaces. Laws 1 and 2 allow relative dating of geomorphic surfaces and help one predict their areal distribution.
3. The shallow subsurface hydrology (soil zone and immediately below) is predictable if one understands the surface materials and geomorphology.

There will be very little emphasis on soil series, soil map units or other classification units, or even soil properties in this text. We emphasize the processes responsible for soil properties at a specific site and how that soil relates to adjacent units. If one understands the genesis of soil materials, then one can predict the site's textural range and possibly the basic soil chemistry. If one knows the geomorphic history of the site, it is then possible to predict the areal extent of soil with properties similar to the one studied. This type of knowledge is important in soil management because it allows one to understand why responses of soil vary.

All current environmental problems involving the soil landscape require a knowledge of stratigraphy, geomorphology and hydrology. Soil is the thin surface veneer of a dynamic, complex three-dimensional system that is active through space and time. Field scientists working in a landscape through the year know by observation where wet areas occur in spring. Someone less familiar with the same landscape may not see the wet areas during infrequent field visits. However, if the individual understands the relationships among stratigraphy, geomorphology and hydrology, he or she can predict the location and duration of the wet areas. Understanding how the processes operate allows one to develop sampling or observation procedures to test predictions accurately.

One purpose of science is to predict. Scientists often need to make predictions where little information is available. We feel strongly that only by using geologic techniques can a soil scientist make reasonable predictions in unfamiliar soil landscapes. Using geologic techniques hones and sharpens the soil scientist's skills in inference and observation. Thus the emphasis throughout the text is on explaining why certain field conditions exist.

FIELD INVESTIGATIONS

The following material is a guide to what soil scientists should look for when conducting field investigations. A competent field person should investigate the following areas of concern.

Stratigraphy

The first question concerns the location of the field site within the stratigraphic column. Are the soil materials parts of a major stratigraphic unit, or are they very local reworked materials that seldom show on geologic maps (Fig. 1.1; Kleiss, 1970)? These questions are important in predicting textural variability where one might expect to find similar materials on the local landscape. What is the depositional environment of these materials? A knowledge of the depositional environment is necessary to understand the potential variability and the areal extent of the unit. The variability and continuity of materials influence the hydrology of the system. In illustration of these ideas, Figure 1.2 shows some textural controls of depositional environment on sediment texture. Loess has a uniform texture and thickness within a local area (Fig. 1.2A). It mantles the pre-loess landscape (Ruhe et al., 1967) so it can occur at any elevation. The depositing stream controls the elevation and areal extent of fluvial sediments so they have a decreasing elevation downstream. Fluvial sediments usually have a fining upward sequence and a wide textural range (Fig. 1.2B) so the textural variability of soils in these materials is large.

Geomorphology

A second question the investigator needs to consider concerns the geomorphology of the area. Is the landscape surface an erosional or a depositional surface? An erosional surface is younger than the immediately underlying materials, but a depositional surface is the same age as the underlying materials. The type of surface determines duration of soil formation and helps predict the variability of soil materials. A depositional surface usually overlies predictable soil materials, though the changes may be abrupt or gradual. For example, nearly level coastal or fluvial plains can have large areas of uniform materials because the depositional environment created similar conditions over a large expanse. By contrast, meandering stream systems may have

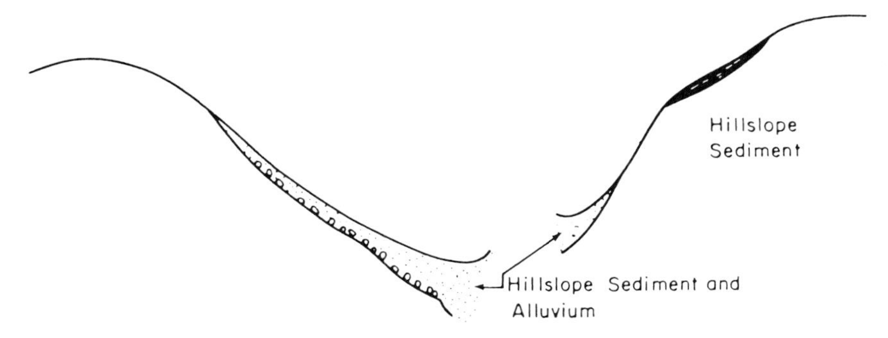

FIGURE 1.1. Hillslope sediment in the North Carolina Piedmont.

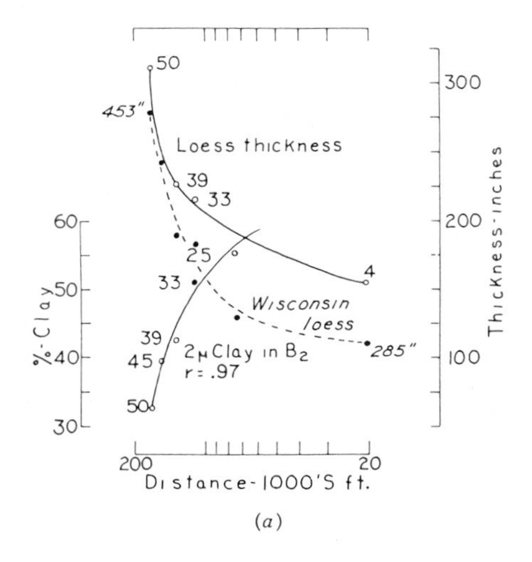

(a)

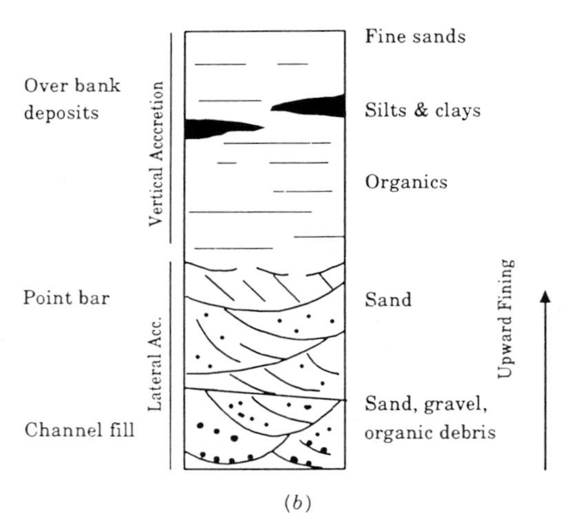

(b)

FIGURE 1.2. A. Clay content and thickness of western Iowa Loess. From Ruhe et al., 1967, Fig. 17, p. 51. B. Textural range of fluvial sediments. Redrawn and modified from Lewis, 1984, Fig. 6, p. 9.

abrupt and distinct vertical and horizontal changes in materials because the energy of deposition changes abruptly.

Soil materials beneath erosional surfaces commonly have abrupt vertical and horizontal changes in materials. Erosional surfaces on fluvial materials expose sediments ranging from clays to gravel (Fig 1.2B) even if the depth of erosion is shallow. By contrast, the textures of loess sheets have little local change. Variability of materials is high in "layer cake" landscapes because erosion exposes several contrasting sediments on the valley slopes.

Erosional Surfaces Some soil scientists relate the age of the soil to the age of the underlying materials. In other words, old materials have thick soils with strong horizonation. This relationship holds only for depositional surfaces. An erosional surface is always younger than the underlying sediments or materials it cuts; therefore, the soils on an erosional surface are younger than the materials from which the soils have developed. Soils on an erosional surface may retain properties imposed by earlier weathering (Fig. 1.3). On erosional surfaces, the site's position on the local landscape is very important (Fig. 1.4). Is the site on an interfluve or on a valley side? If a slope, is it a linear, nose, head or foot slope? Where is the site within the landform? Location is important because soil materials (Figs. 1.2, 1.3) and hydrology vary by position. Sites on nose slopes, shoulders and interfluves may have little or no recognizable materials other than those of the major surface deposit.

Erosional surfaces may be of large areal extent although most of the surface is buried. Even on valley slopes, later sediments derived from upslope bury a

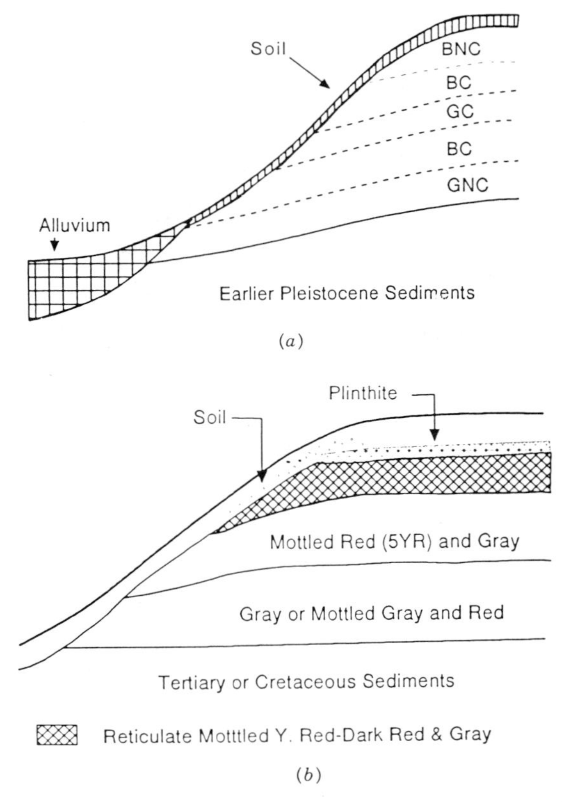

FIGURE 1.3. A. Weathering zones in western Iowa Loess. Redrawn from Ruhe et al., 1956, Fig. 2, p. 346. B. Weathering zones in parts of the Atlantic Coastal Plain.

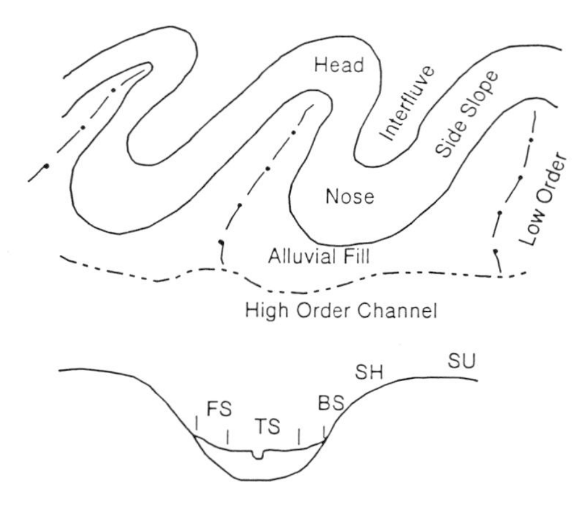

FIGURE 1.4. Cartoon of landform position. Modified from Ruhe, 1969, Fig. 4.5, p. 132. Reprinted by permission from Iowa State University Press.

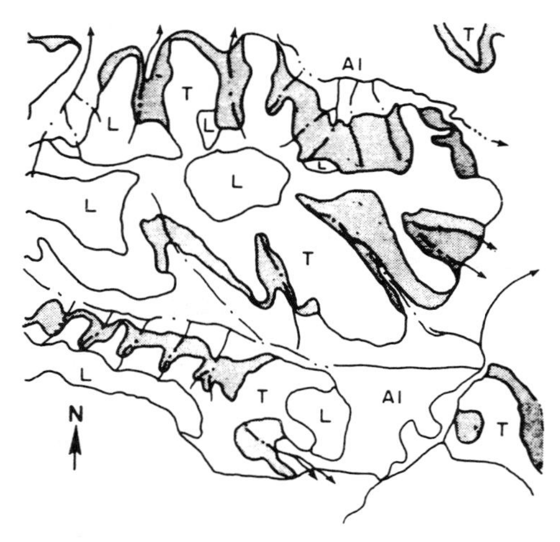

FIGURE 1.5. Areal distribution of hillslope sediment from glacial till in southwestern Iowa. T = till. L = loess. Al = Holocene alluvium. Redrawn from Ruhe et al. 1967, Plate 5.

large part of the erosion surface. On moderate to steep slopes, these locally derived colluvial or water-sorted sediments (hillslope sediments) merge with alluvial fill on the toe slopes. Geologic maps seldom recognize as distinct units these thin materials on slopes (Fig. 1.5).

Many deposits exposed at the surface have weathering zones, or zones of color and other changes, formed by subaerial weathering. In erosional landscapes, the site position in the weathering zones determines the properties (texture and chemistry) of the soil materials at time of exposure (Fig. 1.3). The properties of the preweathered material when first exposed are the starting point for soil formation. The materials may be fresh sediment, saprolite, or a truncated soil profile. Time zero is the period when surface soil horizons started to develop. This development may occur in fresh or preweathered material, or in a truncated soil profile.

Many erosional landscapes have several erosion surfaces present within a small area. This makes it necessary to establish the site's relationship to the larger landscape. Ephemeral upland channels grade to different levels in erosional landscapes with two or more geomorphic surfaces of major areal extent. Figure 1.6 shows how different segments of ephemeral upland channels may be graded to different levels. If all channel segments grade to the same level, a log-log or semilog plot would represent the downstream changes in gradient by a single straight line (Hack, 1957). The upland ephemeral channels shown in Figure 1.6 grade to four different levels. Field recognition of each segment is by the sharp increase in gradient of the lower channel. These channels and their relationships to the adjacent slopes need careful attention.

Hydrology

The shallow subsurface hydrology of the area affects soil properties. The composition of the surface and subsurface materials and the geomorphology of the site control the hydrology. For example, sediment texture (soil texture) has a major control of infiltration and run-off. The subsurface texture and continuity of beds govern the direction and potential rate of water movement. The relief of the area, an expression of the geomorphic history, determines the hy-

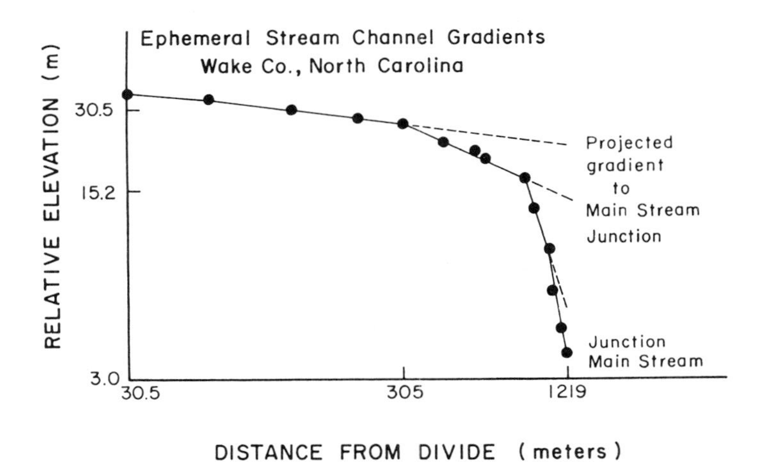

FIGURE 1.6. Ephemeral 1st and 2nd order stream channel slopes in the North Carolina Piedmont. Unpublished data from Schonenberger et al., 1985.

draulic head. The hydraulic head and distance to the outlet govern the rate of water movement within that stratigraphic system. On nearly level surfaces, the distance to the outlet controls how long it takes to drain a site. Drainage is more rapid close to the outlet than at the interstream divide.

On erosional surfaces or depositional surfaces that are not planar, site shape becomes important. Local surface and subsurface flow are converging at concave sites and diverging at convex sites (Zaslavsky & Sinai, 1981). This can produce differences in moisture regimes, transfer and deposition of salts, and in soil drainage. Even under well-drained conditions, surface shape determines how long a site may be moist. Site shape is important in run-off and run-on regimes, and determines which areas are subject to erosion and/or deposition (Young & Mutchler, 1969). Soil properties may exert a control on the local surface hydrology. Site stratigraphy, geomorphology and hydrology determine the kind of soil that develops and affect the rate of soil development.

The Soil System

In all soil investigations it is essential to understand how sediments (soil materials), geomorphology and hydrology interact to produce soils. Without some knowledge of the interactions among materials, surfaces and water movement, the field scientist may be unable to understand soil relationships within the landscape.

Highly detailed transect sampling obtains a large amount of data that can be treated statistically. Grid and transect investigations are useful if we know little about an area. Transect sampling is advantageous in that it has the potential to show progressive changes across a landscape segment and quickly establishes the local stratigraphic and geomorphic relations. However, most field investigations do not have time for grid or detailed transect sampling, and highly detailed grid or traverse sampling and statistical methods alone cannot sort order from chaos. Exhaustive sampling schemes may provide no more insight into internal dynamics than may fewer intensive schemes that consider landforms and surfaces as treatments (Rowe, 1984).

It is necessary to keep in mind the purpose of the investigation. Good field investigations may solve most problems without the aid of laboratory work. When laboratory data are necessary, only a few samples will illustrate the relations uncovered by good field work. Running many samples through the laboratory in the hope that "something will show" is a waste of resources and opportunity.

The sediment or materials from which soils develop determines the soils' initial physical and chemical properties. The depositional environment of the soil materials also influences soil areal variability. The intensity and duration of the processes determine the amount and depth of post-depositional weathering. The age of a surface has no direct influence other than what is done by the given intensity over time. The geomorphic history determines when certain soil processes began or ended. For example, on a broad wet coastal plain surface, organic soils formed until stream dissection created deep drainage of the divides (Daniels et al., 1977). Thus, site history is important in understanding soil development. This history is much more than establishing the age and origin of the surface. We need to know site position on the landscape, when dissection started, and how rapidly dissection proceeded. Site location on the surface controls its hydrology. On erosional surfaces, location is important because it determines whether the site is receiving run-on or is contributing run-off.

Most studies will not require a complete stratigraphic, geomorphic and hydrologic investigation. The scientist needs to understand landscape process to design adequately and interpret correctly field experiments. Poor sampling techniques yield poor statistical results (even if highly significant), but good sampling techniques may provide the data to reveal variability. The major purpose of field experiments should be to produce results that represent the area and allow extrapolation to other locations with confidence.

REFERENCES

Daniels, R.B., E.E. Gamble, W.H. Wheeler, and C.S. Holzhey. (1977). *Soil Sci. Soc. Amer. J.*, 41: 1175–1180.

Hack, J.T. (1957). *Studies of Longitudinal Stream Profiles in Virginia and Maryland.* U.S. Geol. Survey Prof. Paper No. 294.

Jenny, H. (1941). *Factors of Soil Formation*, New York: McGraw.

Kleiss, H.J. (1970). *Soil Sci. Soc. Am. Proc.,* 34: 287–290.

Lewis, E.W. (1984). *Practical Sedimentology*. Stroudsburg, PA: Hutchinson & Ross.

Rowe, J.W. (1984). Forestland classification: Limitation of the use of vegetation. In *Forest Land Classification: Experience, Problems, Perspectives* (pp. 132–147). J.G. Bockheim. Madison: Univ. of Wisconsin.

Ruhe, R.V. (1969). *Quaternary Landscapes in Iowa.* Ames: Iowa State Univ. Press.

Ruhe, R.V., R.C. Prill, and F.F. Riecken. (1955). *Soil Sci. Soc. Am. Proc.*, 19: 345–347.

Ruhe, R.V., R.B. Daniels, and J.G. Cady. (1967). *USDA Tech. Bull. No. 1349.*

Schonenberger, P.J., C.W. Smith, and T. Fox. (1985). Ephemeral channel roll overs (nick points) and their relationship to landscape development in the Piedmont of North Carolina. *Agron. Abs. ASA:* 197.

Young, R.A., and C.K. Mutchler. (1969). *Trans. ASAE,* 12: 231–233, 239.

Zaslavsky, D., and G. Sinai. (1981). *J. Hyd. Div. ASCE*, 107: 37–52.

2 Stratigraphy

INTRODUCTION

Our purpose in this section is to acquaint the reader with the principles used by stratigraphers. These principles are important in understanding and predicting the properties and distribution of soil materials. Stratigraphy also is the foundation for most geomorphic investigations.

Many early soil scientists were geologists who placed considerable emphasis upon the origin of soil materials. Their early literature refers frequently to soil formed in alluvium, till, or other materials. The modern emphasis has dramatically shifted, however, from interest in the kind of soil material to classification of soils by their morphological and chemical properties.

Not everyone agrees with the change in emphasis forced by taxonomy from materials to morphology because most realize that problems would occur. We suggest that serious students become acquainted with the older literature to understand the problems produced by the current emphasis on morphology alone. Simonson's (1959) generalized theory of soil genesis emphasizes the processes that develop soil horizons. He also points out that soil genesis has two steps: (1) the accumulation of parent materials, and (2) the differentiation of profile horizons. Cline (1961) believed that the basic ideas of geology, rock and minerals and their transformations, are keystones of soil genesis. He emphasized that geomorphology is the foundation of our interpretations of past events. Knox (1965) argued that the pedon is an arbitrary unit and suggested that soil landscape units should be the basis of classification.

If the origin of soil materials is known, it is easier to predict soil distribution and variability. Homogeneity of soil materials will be a function of source, transport, and depositional environment. Knowledge of these processes is important when designing field experiments. The experimental design should include intensity, depth, pattern and number of samples required for robust statistical analysis. Experimental design, in turn, should include consideration of the relative uniformity of the soil materials.

Soil Variability

Much of the soil variability on a landscape results from material variability. For example, the most uniform soil materials known are thick loess sheets. The loess sheets of southwestern Iowa have a west to east soil association of Ida-Monona-Hamburg, Marshall-Monona, Marshall, Sharpsburg, etc. (Fenton et al., 1971). Within a soil association area it is easy to predict what soils occur within the landscape. The textures within a loess soil association are very uniform locally. Textures grade from coarsest nearest the source to finest downwind. One can estimate the clay content of the Bt horizon within 2 to 4 percent by knowing the geographic location within the association. This does not imply that different soils do not occur within the thick loess sheets, because they do. Soil properties and location of specific soils on the local loessial landscape are predictable.

By contrast, predictions of textures within a small flood plain are more difficult. Predictions are poor because flood plain sediments have abrupt vertical and horizontal textural changes. Coarse point bar deposits occur at the outside bends in the channel. Fine sand to silty natural levees a few meters wide may parallel the channel. Overbank deposits away from the channel may grade laterally into clayey sediments. Most fluvial deposits have a fining upward sequence (Fig. 1.2), but clayey channel deposits can abruptly overlie coarse channel lag deposits. Considerable horizontal changes in textures can occur in short distances in areas with thick fluvial sediments. For more detail on the areal distribution of textures in a fluvial landscape, see Chapter 4.

Landforms on unmodified depositional surfaces or plains have a narrow range of sediment textures. This is particularly true in young landscapes with little truncation and reworking of surface materials, or within the stable portions of older landscapes.

Textural changes in dissected fluvial sediments are difficult to predict because the present landform has little or no relationship to the process responsible for the sediment texture. In portions of the Ridge and Valley province of Tennessee, for example, soil profiles in the highest ridges commonly contain stratified rounded lag gravel (Springer & Elder, 1980). These alluvial deposits are relicts from times when drainage base levels were much higher than present.

One purpose of science is to use current knowledge to predict what will happen under a given set of circumstances or what one will find within a specific environment. Field soil scientists must continually predict what they will find. Successful prediction of soil distributions within a landscape depends upon the development and refinement of one's ideas (or model) of that soil landscape. The most accurate idea (model) is one based upon local geomorphic and stratigraphic conditions. Most experienced individuals are accurate within their work areas; however, experience in one area may be of little value in prediction in another area. The worker needs a knowledge of the textural characteristics of the materials by depositional environment. Order exists in all depositional systems. Someone who knows where he or she is within that system has a reasonable chance of making accurate predictions. A person with such knowledge also can better understand the local soil variability.

Data Analysis Many soil scientists now use "geostatistics" to predict the variability of a particular area or soil (Nielsen & Bouma, 1985; Trangmar et al.,

1985). We submit that the term "geostatistics" is a broad generalization having no more practical meaning than terms such as "tropical soils" or "deciduous forest ecosystems." Some workers espousing geostatistical techniques apparently have little knowledge of soil materials and processes; grid sampling and variagrams are impressive when presented, but frequently they tell us less than we can learn from geologic maps and facies descriptions.

Experimental designs and sampling techniques that do not consider geomorphology, stratigraphy, and soil water movement as soil-forming treatments are not likely to reveal information transferable to other landscapes. The idea that soil variability is random and that we cannot predict it within reason is usually false. Much of the variability is sequential and predictable, but we can never be completely confident in our predictions. Scale of observation is important. We can have reasonable predictions for average clay content throughout a field, but can not predict bulk density changes within a small plot over time. If we cannot predict variability within a reasonable range, we do not understand the system. If we do not understand the system, we cannot accurately interpret statistical results or predict the results from environmental changes.

A review of Davis's (1986) treatise of statistical methods for geologists should be mandatory for those embracing or considering geostatistical analyses. Davis has stated his case eloquently, and begins by saying that the science of geology

> has benefitted more than it has contributed to the exchange of quantitative techniques. . . . data are derived from scattered drill-holes that pierce successive stratigraphic horizons. The elevation of the top of a horizon measured in one of these holes constitutes a single observation. Obviously, an infinite number of measurements of the top of this horizon could be made if we drilled unlimited numbers of holes. This cannot be done . . . most geologists must take their observation where they can. Paleontologists must be content with the fossils they can glean from the outcrop; those buried in the subsurface are forever beyond their reach. (pp. 5–6)

After his discussion of the reasons for using statistics to analyze geologic data, Davis continues with a note of caution.

> In this text, we will concern ourselves with those less tractable situations where the sample design (either by chance or ignorance) is beyond our control. However, be warned that an uncontrolled experiment (i.e., one in which the investigator has no influence over where observations are taken) usually takes us outside the realm of classical statistics. This is the area "quasistatistics" or "protostatistics" where the assumptions of formal statistics cannot safely be made . . . and the best we can hope from our procedures is help in what ultimately must be a human decision. A quantitative approach to geology requires something more profound than a headlong rush into the field armed with a personal computer. The geologist must be aware of the nature of . . . the systems in which the measurements are made. (1986, pp. 6–7)

Davis feels that perhaps the most positive benefit of the use of mathematical analyses in geology is

> that a quantitative approach to geology can yield a fruitful return to an investigator . . . in the insights gained through the critical examination of phenomena, which is required by a quantitative method. *The gathering of data of sufficient quality and quantity to be useful in numerical analysis forces a closer familiarity with the objects of study than might otherwise be obtained. . . . At the same time, the discipline necessary to perform quantitative research will hasten the growth and maturity of the scientist* [Italics added]. (1986, pp. 4–5)

We heartily endorse Davis's philosophy. Numerical analyses help us gain further insight into the systems of interest. They should supplement, not replace, *careful field observation*. Soil scientists should be careful when using statistical procedures. Considerable understanding of statistical theory is necessary to generate relevant results. As Davis (1986) points out, these results are not dogma, but rather are subject to user interpretation. We should remember that random sampling and statistical analyses, no matter how intensive the sampling or how powerful the statistical procedures, cannot replace careful, thorough field research. We must incorporate into the experimental design all we know about the soil materials.

STRATIGRAPHIC PRINCIPLES AND CONCEPTS

We do not use stratigraphy in this text in the classical sense. Our interest is in the soil materials and their distribution. Emphasis will be on surface materials, their origin and properties, and on the weathering profile. The weathering profile may or may not be thicker than the solum. The material may be a formal stratigraphic unit or the seldom recognized hillslope sediment.

Definitions

Stratigraphy is that branch of geology dealing with the nature and distribution of layered (stratified) rocks of the earth's crust. A stratigrapher investigates the age, areal extent, properties, and depositional environment of stratified rocks. The stratigraphic column is a fundamental part of each geology report on an area. The column names the major stratigraphic units and places them in a relative or absolute time sequence (Fig. 2.1). The composite stratigraphic column in Figure 2.1 is only an attempt to show where each mapped unit fits into an age sequence. It does not mean that the units always (although they frequently do) directly overlie each other on the landscape. They may be kilometers apart. Several different units, groups, formations, members or beds are in the column, but the formation is the fundamental stratigraphic unit. A formation is a mappable unit. It may cross time lines, environments and lithologies, or have very narrow ranges of the above. A formation is concep-

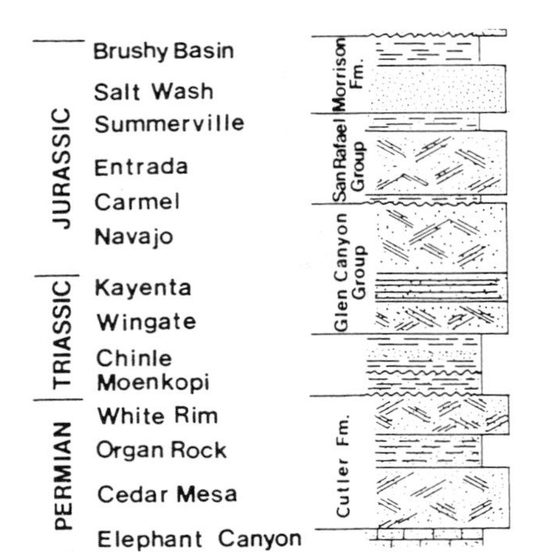

FIGURE 2.1. Typical stratigraphic column. Redrawn from Jackson and Pollard (1988, Fig 5, p. 121). Reprinted by permission from Geological Society of America and M.D. Jackson.

tually very similar to a soil mapping unit. Delineations of the same unit contain similar characteristics and occur in predictable positions.

Stratigraphy has a strict set of rules on how to identify, name and classify the units mapped. The code is in the 1983 *American Association of Petroleum Geologists* Bulletin (AAPG). It is the taxonomy of stratigraphers but probably has more flexibility than *Soil Taxonomy* (Soil Survey Staff, 1975).

Superposition

The law of superposition states that younger beds overlie older beds, unless overturned or thrust-faulted (Fig. 2.2). This law is essential to help decide the relative ages of various mapped units. Superposition is a major field tool when fossils or other datable materials are absent in surface deposits. It is a useful dating method during field work and helps solve many geomorphic problems. Note that the number sequence of sediments in Figures 2.2–2.7 does not always follow standard practice. This is done to emphasize relationships rather than numbers.

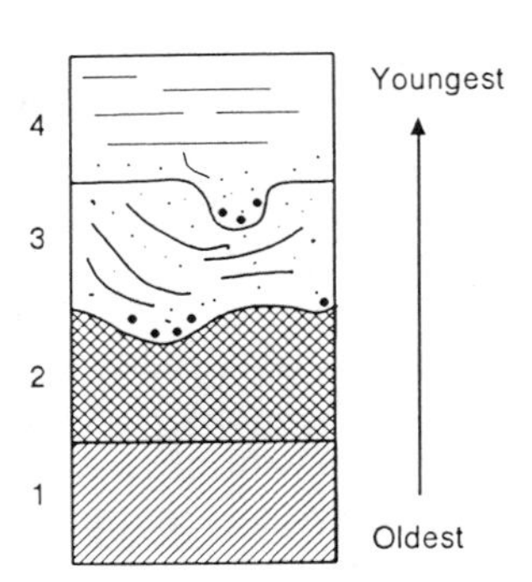

FIGURE 2.2. A cartoon of superposition.

Relative Dating

Figure 2.3 illustrates a sequence of terraces overlying bedrock. The relative age sequence is 1>2>3>4. Unit 4 is inset below 3 and therefore is younger than 3. This shows one can date beds or materials that do not overlie one another. Their positions in relation to each other are also important. In Figure 2.3 we infer the age relations of fluvial units 2, 3 and 4 from surface form alone. But interpretations must be cautious because stratigraphic proof is lacking. The sediments and their areal relations are important, not the surface form, although form may give a clue about origin.

The first idea one would test is that 2, 3 and 4 are different units. Figure 2.4 shows a stepped surface sequence mapped as one unit or formation. We only know that 2 is younger than 1. There are several possible explanations for the stepped sequence. It is possible that exposures are lacking and lithologies across the scarps or flexures are too similar to separate different sediments. Each scarp or step could represent a different depositional unit, or they could be erosion surfaces cut into unit 2. If the scarps are of marine origin, they could form one transgressive-regressive unit. A pause in the drop in sea level produced each scarp. The upper part of the unit is younger toward the lower step with possible large time discontinuities at the scarps.

Figure 2.5 illustrates the age relations among tills on broad till plains separated by end moraines. In this situation, the materials beneath the lower steps are older, with the sequence being 1>2>3 because of superposition. However, one should be careful in using form alone when dating sediments underlying stepped surfaces; if the units in Figure 2.5 were fluvial or marine terraces, the age relations are opposite those from using form alone (Fig. 2.3).

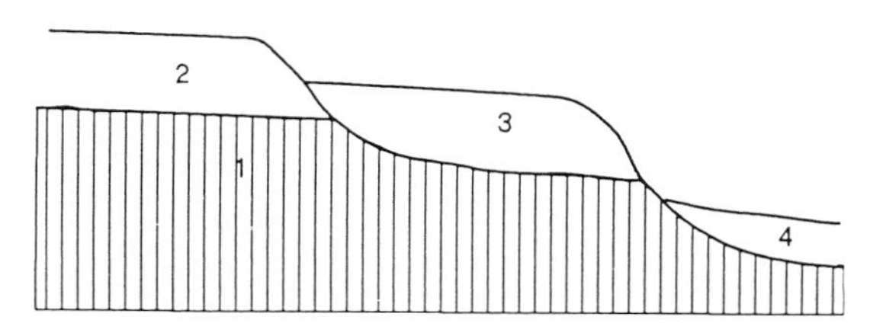

FIGURE 2.3. Ages of river terraces.

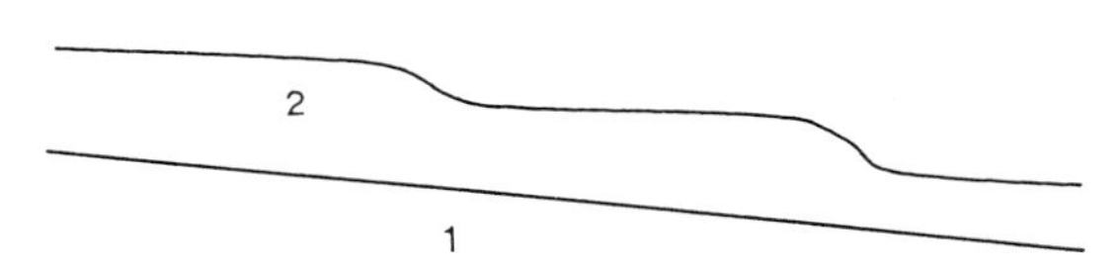

FIGURE 2.4. A stepped sequence of terrace surfaces without a distinct change in sediments.

The depositional environment, not the surface expression, determines the age relations of these units.

Scarped and mantled landscapes (Fig. 2.6A) are another example of the problems involved in dating materials by position only. The relative ages are 1>2>3>5>4>6 in Figure 2.6A. Unit 4 is loess or eolian sand. Unit 6 truncates and is inset below 4 so it is younger than 4. Eolian sand in Figure 2.6B mantles units 1 and 2 and interfingers with unit 5. The relative age relations are 1>2>3=5 in Figure 2.6B. The upper part of 5 is younger than 3, but the units are within the same depositional period. The relationships shown are simple and easily recognized in the field when one knows the surface stratigraphy. Usually the actual relationships will be more complex, but the fundamental principles of relative age determination apply.

Superposition states that a bed is younger than any other bed it overlies. This means a sediment is younger than any other sediment it overlies or truncates. The simplest example of truncation is a cross-cutting fault or dike system (Fig. 2.7). In Figure 2.7, 2 is older than 1, because 1 truncates 2 and there apparently has been some displacement. The reader is urged to establish a

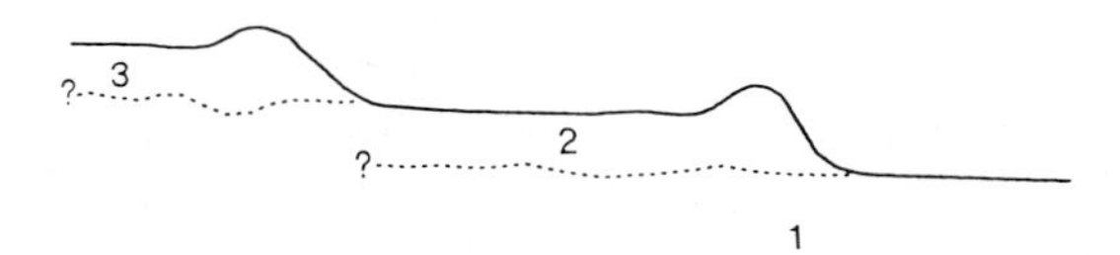

FIGURE 2.5. Age relations among ground moraines and minor moraines.

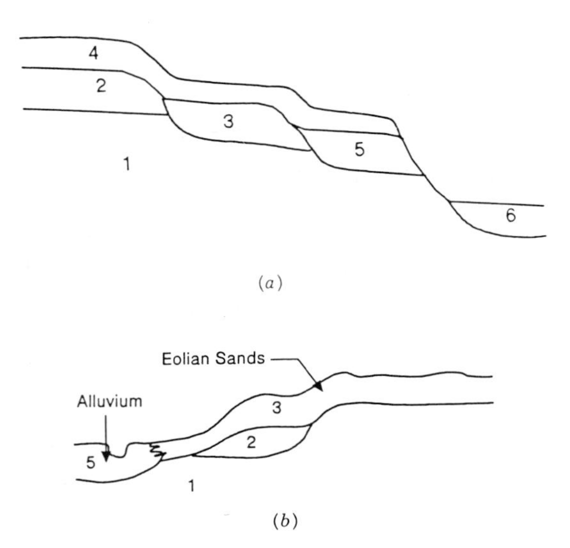

FIGURE 2.6. A. Age relations among loess-mantled terraces. B. Age relations in an eolian sand-mantled landscape.

relative age sequence for Figure 2.8. The same criteria apply to the areal distribution of sediments.

Terraces Figure 2.9 is a plan view of a valley terrace system. Convention numbers terraces from highest to lowest, so T1 is the highest and T2 the lowest terrace. If the different levels in Figure 2.9 represent different periods of deposition, T2 is younger than T1. Terrace T2 truncates T1, and its surface is lower than T1. The same relationship exists between T2 and T3. Although the above inferences may be correct in river valleys with distinct terrace systems, proof is lacking. It is possible that T2 and T3 are the same age and the scarp is a fault line scarp. If faulting is responsible for the inset relationship of T2 and

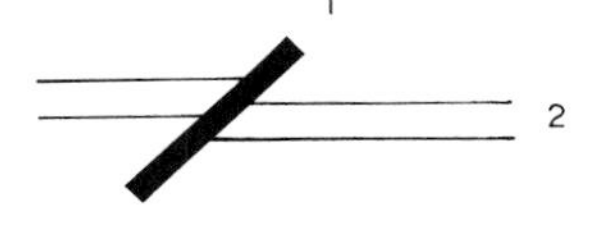

FIGURE 2.7. Relative age relations of cross-cutting dikes.

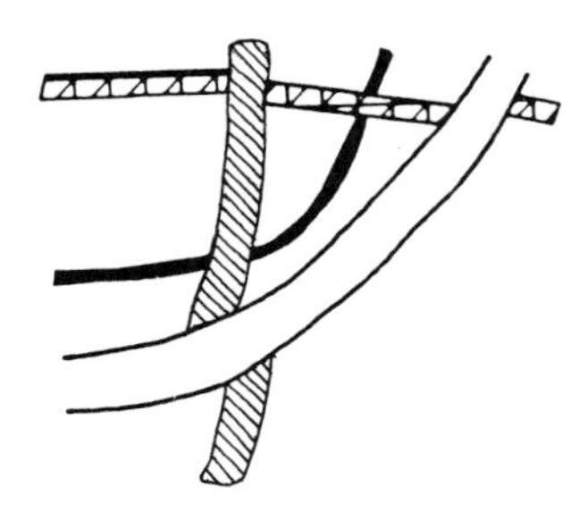

FIGURE 2.8. Age relations of complex cross-cutting dikes.

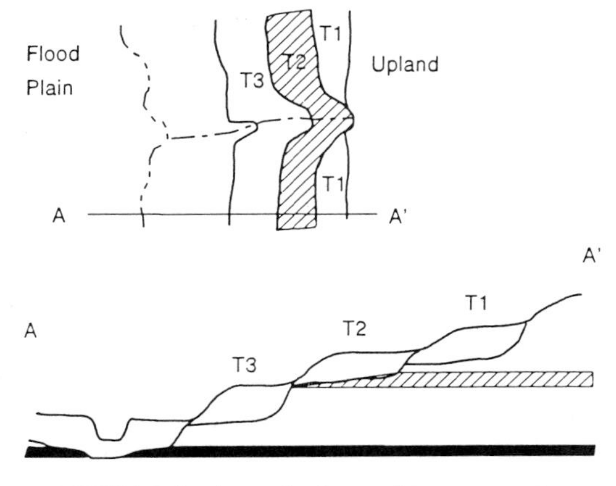

FIGURE 2.9. Age relations of terrace systems.

T3, then only T2 would not occur within the tributary valleys. Because T3 is inset below T2 in the tributary valleys, other processes explain the relationships shown. It is also possible that the terraces are *straths* with little or no alluvial fill beneath the surface. Another possibility is that each terrace surface is an erosion surface cut into a preexisting alluvial fill.

Madole's work (1988) on the Meers fault is an excellent example of the importance of stratigraphic proof required to date terraces. The landscape in Figure 2.10 has two terraces of different ages. Other features produced by the faulting, which involved throw and lateral displacement, dammed small gullies and ponded fine-grain alluvium upslope from the scarp (Crone & Luza, 1990). The dammed gullies are easily visible on aerial photographs.

Without the stratigraphic evidence of faulting, the interpretation would be that the stream trenched and removed most or all the Browns Creek alluvium, and then deposited the late Holocene (fault-related alluvium of Madole, 1988) alluvium. If the fault-related fan alluvium thins downstream, the Browns Creek alluvium would come to the surface. The Browns Creek is then a downstream facies change of the late Holocene fault-related alluvium. The surface and age of the lower terrace would be the same whether it was fault-related fan alluvium or the outcrop of the Browns Creek alluvium downstream from the fault. One also would have difficulty in interpreting and explaining the low terrace-like feature of shale bedrock that is about the same elevation as the terrace or lower surface composed of the fault-related fan alluvium (Fig 2.10). Also, it would be difficult to establish the age of the fine-grained alluvium dammed by the lateral displacement. One must always check the first impression of form against the sediments that underlie that form.

The final proof that the terraces are episodes of valley cutting and filling is in the cross section of Figure 2.9. Each terrace sediment is inset below the higher. Marker beds beneath the upland continue beneath the scarps without interruption. The terrace system developed from successively lower base levels. The depositional environment could range from fluvial to marine. Much can be inferred from surface form, but the final proof is in the detailed cross-sections.

Marine and Glacial Sediments Figures 2.11, 2.12 and 2.13 show how to use truncation of older materials to place sediments in a relative time sequence. These are common distributions of materials on coastal plain and glaciated landscapes. The relations are much more complex in mountainous terrain, but the same principles apply.

In Figure 2.11 the Surry scarp truncates units 1 & 2 and the later flood plain deposits truncate the scarp. We cannot accurately date the fine-textured member (1) and the sandy member (2) with available data. Figure 2.12 illustrates how truncation and geographic distribution of minor moraines help establish relative ages. The age sequence of the materials is $2>3=1$. Units 3 and 1

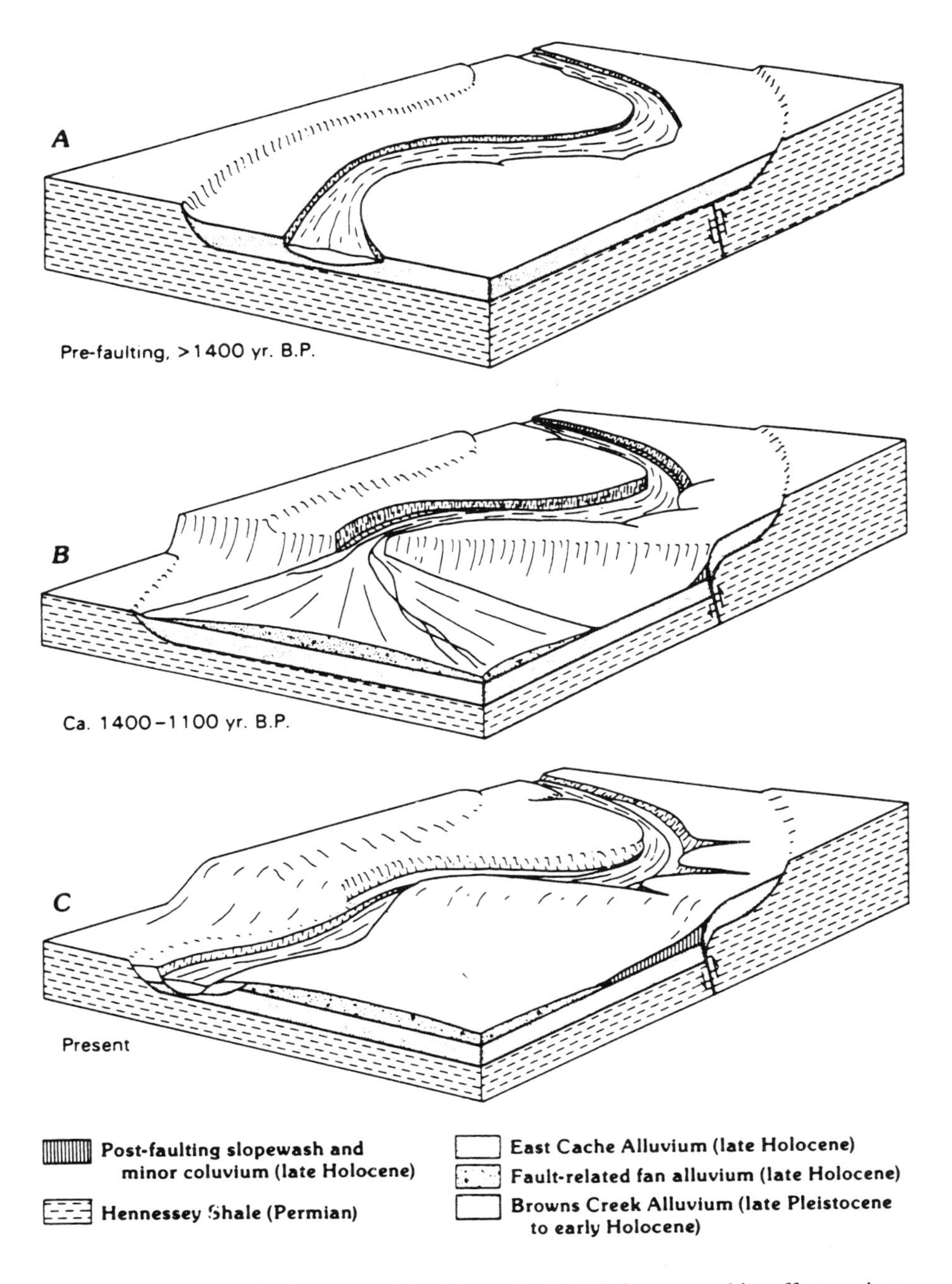

FIGURE 2.10. Block diagrams of the Meers fault, Oklahoma, and its affect on interpretations of sediments and landforms. From Madole, 1988, Fig. 7, p. 398. Reprinted by permission from Geological Society of America and R.F. Madole.

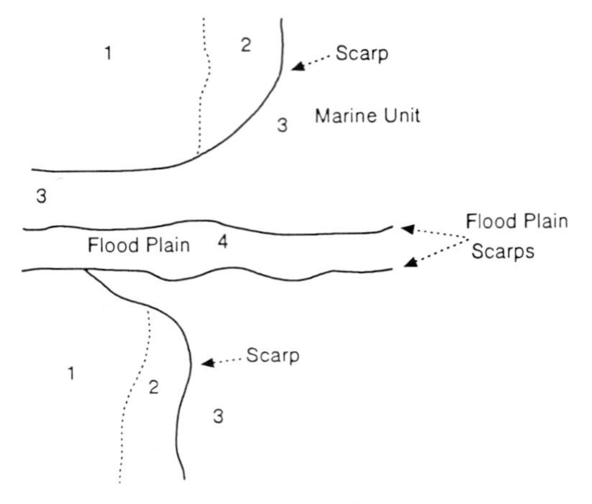

FIGURE 2.11. Areal relations of sediments in Coastal Plains.

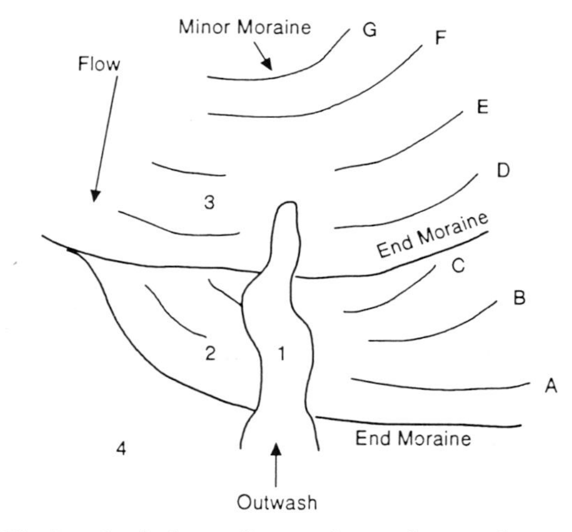

FIGURE 2.12. Areal relations of ground moraines and associated deposits.

truncate unit 2 and the outwash plain is the same age as 3. The age sequence of the minor moraines is (A) the oldest and (I) the youngest.

Loess-mantled landscapes are complex but interesting. The age sequence in Figure 2.13 is 2=>T1>3>1=4=5. We will assume that the Missouri Valley alluvium is Holocene and that till overrides the loess. Terrace T1 is a wide nearly level loess-mantled remnant with about ½ the normal loess thickness in the area. From this information, we can show that units 4 and 5 probably are the same or similar age. The upper half of the upland loess mantle buries the T1 alluvium. Thus the T1 alluvium may be the same age as the lower half of the loess. If the alluvial surface is an erosion surface, the T1 alluvium can be older than the loess. The other relationships are straightforward.

Mud Flows Figure 2.14 shows mud flows on the valley slope. They overlie the same material, and the age relation between the flows is unknown because one neither overlies nor truncates the other.

Geochemical Dating Methods

Several methods of dating sediments are available, and each requires special techniques and skills. Radiocarbon dating probably is most useful for soil studies, but at times other methods are better. Table 2.1 shows some geochemical dating methods used, their range in age and some associated problems.

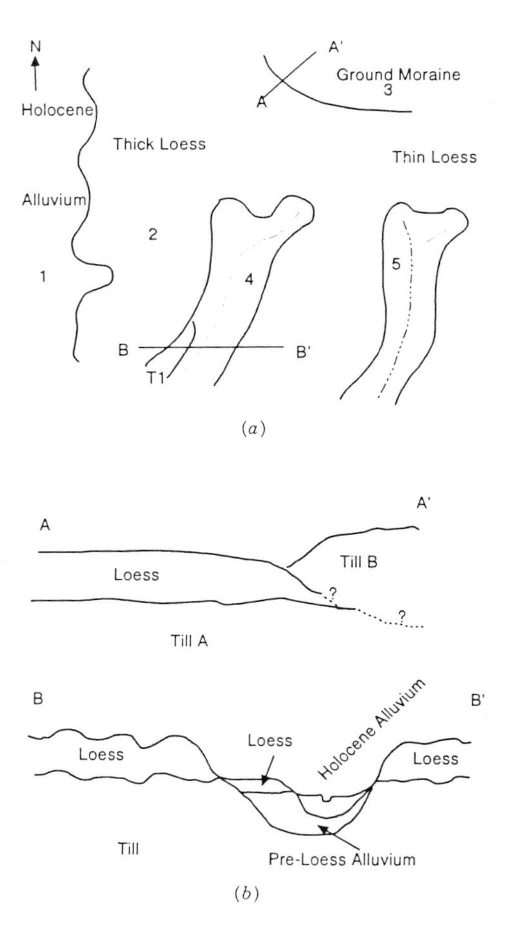

FIGURE 2.13. Age relations in loess-mantled topography.

TABLE 2.1. Apparent Age Techniques

Method	Approximate Time Span (Years)	Data Material and Problems
Cesium 137	30–50	Soils and sediments. Leaching, associated with clays.
Lead 210	<150	Sediments. Industrial contamination.
Carbon 14	50,000	Wood, peat, $CaCO_3$. Contamination, poor sampling.
Thermoluminescence	5×10^4 quartz 1×10^6 feldspar 1×10^5 ocean sediments	Eolian, ocean, terrestrial sediments, tephra, buried soils, pottery. Tedious laboratory techniques.
Potential for extending range to several 100 thousand years		
Fission-track	$>1\times10^5$	Tephra, zircon, landscape evolution (heating). Grain-discrete; highly detailed laboratory procedures.
Beryllium 10	$>2\times10^5$	Soils, rain, freshwater sediments, tektites. Loss by erosion, leaching.
Paleomagnetism, magnetic Susceptibility		Must be correlated with other methods. Useful for tills, loess, waterlaid sediments, mudflows and some iron-rich post-depositional weathering features. Requires silt to clay rich materials and supporting dating by other procedures.
K40 Ar	10^4—oldest	^{40}K abundant in most rock-forming minerals. Restricted to igneous and metamorphic rocks. Glauconite the only sedimentary material.
Other Relative Dating Techniques		
Desert Varnish		
Dendochronology		
Lichenometry		
Pitting of Boulders		

Compiled from several sources given in the references and bibliography.

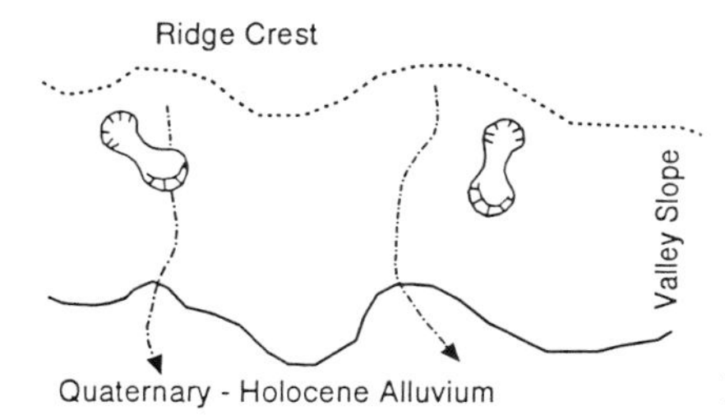

FIGURE 2.14. Areal relations of mud flows.

Most of the methods require detailed and complicated laboratory techniques and often need strict sampling procedures to produce valid data.

REFERENCES AND BIBLIOGRAPHY

AAPG, 1983. North American Stratigraphic Code, North American commission on stratigraphic nomenclature. *Amer. Assoc. Petroleum Geol. Bull.*, 67: 841–875.

Aitken, M.J. (1985). *Thermoluminescence Dating.* New York: Academic Press.

Aller, R.C. and J.K. Cochran. (1976). *Earth Planet Science Letter*, 20: 37–50.

Berger, G.W. (1984). *Canadian J. Earth Sciences*, 21: 1393–1399.

Berger, R. and H.E. Suess. (Eds.) (1979). *Radiocarbon Dating.* Berkeley: University of California Press.

Birkeland, P.W., S.M. Colman, R.M. Burke, R.R. Shroba, and T.C. Meierding. (1979). *Geology*, 7: 532–536.

Burke, R.N. and P.W. Birkeland. (1979). *Quaternary Research*, 11: 21–51.

Cline, M.G. (1961). *Soil Sci. Soc. Amer. Proc.*, 25: 442–446.

Cohee, G.V., M.F. Glaessner, and H.D. Hedberg. (Eds.) (1978). *The Geologic Time Scale.* AAPG Studies in Geology, No. 6.

Crone, A.J. and K.V. Luza. (1990). *Geol. Soc. Amer. Bull.*, 102: 1–17.

Currie, L.A. (1982). *Nuclear and Chemical Dating Techniques.* Washington, DC: American Chemical Society.

Dalrymple, G.B. and M.A. Langhere. (1969). *Potassium-Argon Dating.* San Francisco: W.H. Freeman.

Davis, J.C. (1986). *Statistics and Data Analysis in Geology*, 2nd ed. New York: Wiley.

de Vernal, A., C. Causse, C. Hillaice-Marcel, R.J. Mott, and S. Occhietti. (1986). *Geology*, 14: 554–557.

Dorn, R.I., and T.M. Oberlander. (1982). *Progress in Physical Geography*, 6: 317–367.

Dreimanis, A., G. Hutt, A. Raukas, and P.W. Whippey. (1978). *Geoscience Canada*, 5: 55–60.

Easterbrook, D.J. (Ed.) (1988). *Dating Quaternary Sediments.* Geol. Soc. Amer. Special Paper, No. 227.

Fenton, T.E., E.R. Duncan, W.D. Shrader, and L.C. Dumenil. (1971). *Productivity Levels of Some Iowa Soils.* Iowa. Ag. and Home Econ. Exp. Sta. Special Report, No. 66.

Hall, R.D. and R.E. Martin. (1986). The etching of hornblende in the matrix of tills and periglacial deposits. In *Rates of Chemical Weathering of Rocks and Minerals* (pp. 101–128). Ed. by S.M. Coleman and D.P. Dethier. Orlando, FL: Academic Press.

Hall, R.D. and D. Michaud. (1988). *Geol. Soc. Amer. Bull.*, 100: 458–467.

Huddleson, J.H. and F.F. Riecken. (1973). *Soil Sci. Soc. Amer. Proc.*, 37: 264–270.

Jackson, M.D. and D.D. Pollard. (1988). *Geol. Soc. Amer. Bull.*, 100: 117–139.

Kachanoski, G. (1985). *Estimating Soil Erosion in Ontario Using Soil 137-Cesium*. [Land Resource Science 1985 annual report.] Guelph, Ont: University of Guelph.

Knox, E.G. (1965). *Soil Sci. Soc. Amer. Proc.*, 29: 79–84.

Krishanasami, S., L.K. Benniner, R.C. Aller, and K.L. Van Damm. (1980). *Earth Planet Science Letter*, 47: 307–318.

Kula, G., F Heller, L.X. Ming, X.T. Chun, L.T. Sheng, and A.Z.Sheng (1988). *Geology*, 16: 811–814.

Locke, W.W. (1986). Rates of hornblende etching in soils on glacial deposits, Baffin Island, Canada. In *Rates of Chemical Weathering of Rocks and Minerals* (pp. 129–145). Ed. by S.M. Coleman and D.P. Dethier. Orlando, FL: Academic Press.

Madole, R.F. (1988). *Geol. Soc. Amer. Bull.*, 100: 392–401.

Nielsen, D.R. and J. Bouma. (Eds.) (1984, November). *Soil Spatial Variability*, Proceedings of the Workshop of ISSS and SSSA. Las Vegas, NY. Pudoc Wageningen (1985).

Pavich, M.J., L. Brown, J.N. Valette-Silver, J. Klein, and R. Middleton. (1985). *Geology*, 13: 39–41.

Pavich, J.J., L. Brown, J. Harden, J. Klein, and R. Middleton. (1986). *Geochemica et Cosmachemica Acta.*, 50: 1727–1735.

Rice, D.L. (1986). *J. Marine Research*, 44: 149–184.

Ridge, J.C., W.J. Brennan, and E.H. Muller. (1990). *Geol. Soc. Amer. Bull.*, 102: 26–44.

Simonson, R.W. (1959). *Soil Sci. Soc. Amer. Proc.*, 23: 152–156.

Singhvi, A.K. and V. Mejdahl. (1985). *Nuclear Tracks*, 10: 137–162.

Singhvi, A.K., Y.P. Sharma, and D.P. Agrawal. (1982). *Nature*, 195: 313–315.

Soil Survey Staff. (1975). *Soil Taxonomy.* [Agric. Handbook, No. 436.] Washington, DC: U.S. Govt. Printing Office.

Springer, M.E. and J.A. Elder. (1980). *Soils of Tennessee.* Univ. of Tennessee Agric. Exp. Sta. Bull, No. 596.

Stewart, A.J., D.H. Blake, and C.D. Ollier. (1986). *Science*, 233: 758–761.

Trangmar, B.B., R.S. Yost, and G. Uehara. (1985). *Advan. Agron.* 38: 45–95.

Wintle, A.G. (1973). *Nature*, 245: 143–144.

Wintle, A.G. (1982). *Soil Sci.*, 134: 164–170.

Wintle, A.G. (1987). Thermoluminescence dating of loess. In *Loess and Environment* (pp. 103–116). Ed. by M. Pecsi. Catena Supplement 9.

Wintle, A.G. and D.J. Huntley. (1982). *Quaternary Science Reviews*, 1: 31–52.

3 Textural Characteristics of Soil Materials

INTRODUCTION

Stratigraphy is a very useful science in soil studies. Stratigraphic concepts help one establish the sequence of sediments, their areal distribution, relative or absolute age and probable depositional environment. Classical stratigraphy and soil classification have a conceptual organization that provides a framework, but little detail. Logic is the base of any classification scheme and does not require a knowledge of the distribution of properties of the objects being classified. The early development of soil taxonomy (Soil Survey Staff, 1975) included a philosophical debate of the most preferable taxonomic scheme (Cline, 1949, 1963; Smith, 1963).

Soil scientists need the information found in facies descriptions. These descriptions give the textural ranges of soil materials within the facies. Modern research in sedimentology is useful to soil scientists interested in the areal and vertical variability of soil materials. Field soil scientists with a knowledge of sedimentation processes can refine their conceptual (predictive) model of soils in the landscape.

For example, some Coastal Plain soils in North Carolina are subject to manganese deficiency when limed to a pH of near 7 (Miner et al., 1986). There are large differences in soil test manganese between the Coastal Plain and Piedmont (Table 3.1). The soil test manganese levels in the Coastal Plain counties are low because barrier and marine sediments are naturally low in manganese. Piedmont soils from mafic and felsic bedrock have high manganese levels (Table 3.1).

One would predict from these data that manganese is low in soils on terraces of streams draining only the Coastal Plain. Soils on terraces in river valleys draining both Piedmont and Coastal Plain probably test high in manganese. These soils should not have manganese deficiencies if limed to a pH above 7.

Other depositional environments also have specific characteristics that are important to soil use. This discussion will emphasize textural properties. Abundant textural data are available in the geologic literature, and many soil properties have a high correlation with soil texture. Figure 3.1 gives the phi units used by geologists and the millimeter (mm) scale used by soil scientists.

TABLE 3.1. Manganese Soil Test Levels

	Soil Test Values (% of Samples)				
	0–10	11–25	26–50	51–100	>100
			County		
			Coastal Plain		
Pender	2.93	22.84	44.21	22.80	7.22
New Hanover	10.77	19.41	28.34	24.76	16.77[a]
			Piedmont		
Rockingham	0.0	0.58	1.47	7.76	90.20
Cabarrus	0.14[b]	0.14[b]	0.28[c]	2.24	97.20

[a] Includes large number of samples from lawns and gardens.
[b] One sample in 714.
[c] Two samples in 714.

Source: Unpublished data, North Carolina Department of Agriculture Soil Testing Laboratory, Raleigh, North Carolina, 1985.

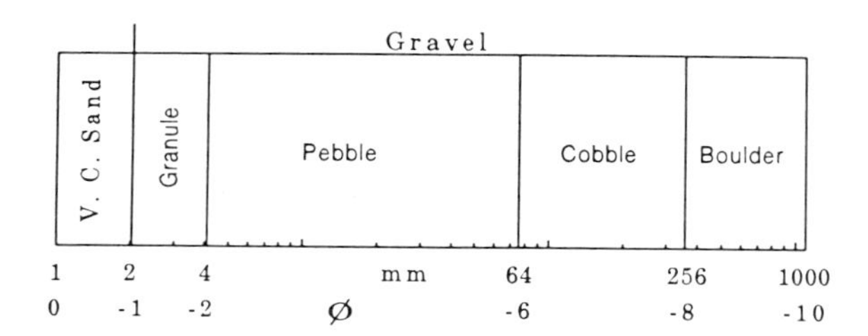

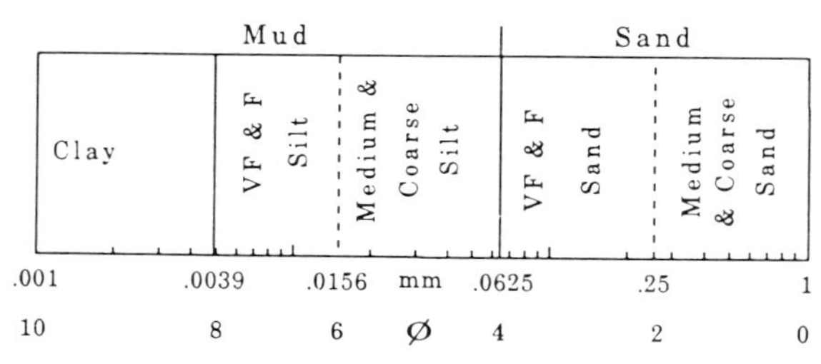

FIGURE 3.1. Common scales used by geologists, engineers and soil scientists. From Lewis, 1984, Fig. 37, p. 59.

TERRESTRIAL ENVIRONMENTS

Lewis (1984) divides depositional environments into non-marine, transitional and marine units. We will confine our discussion to the non-marine, transitional and shallow marine environments. These environments are common soil materials and are most likely to have remnants of depositional surfaces. Non-marine environments include terrestrial and lacustrine units. Terrestrial environments include glacial (moraines and outwash), eolian, river flood plains, river channels, and alluvial fans and plains.

Glacial Drift

Glaciers covered nearly 30% of the earth in the last 3 million years. Loess deposits retain records of 17 glacial advances (Chorley et al., 1985). Glacial drift covers about 8 percent of the earth's surface above sea level and about 25 percent of the North American continent (Ritter, 1986). Drift ranges in thickness from a few meters to several hundred meters in buried valleys, but is about 10 to 60 meters (m) thick in the central United States (Ritter, 1986). In the Northern Hemisphere, glacial deposits can be a major soil material even in deeply eroded landscapes.

Till probably is one of the most variable sediments (Goldthwait, 1972). Bedrock source, transport distance and depositional subenvironment all influence till sheet texture. The bedrock scoured by the glacier controls the till composition (Dreimanis & Vagners, 1972). Ritter (1986) notes that the great variety of glacial drift and the complex interrelationships that exist between the different types intrigue geologists. He states that these complex relations arise

> because (1) drift may be deposited from mediums that contain vastly different amounts of water; (2) deposition occurs beneath, within, or on top of the ice, at the glacier margins, in bodies of standing water, or in fluvial settings far from the glacier, the debris being transported there by streams rising in the ice mass itself; (3) the depositional sites and environments and the drift composition all change with time because glaciers themselves are not constant in their properties or fixed in their position; and (4) the glacier may be active or stagnate. (pp. 394)

Table 3.2 is one of several classifications of glacial drift. Lawson's publications (1981a, 1981b) should be consulted for more detail. Ritter's separation of drift into nonstratified (till) and stratified material (fluvoglacial) is a good starting point. As the names imply, nonstratified drift originates directly from the glacial ice and lacks stratification. Glacial water transports and sorts stratified drift. Several subenvironments (Fig. 3.2) occur within the two way classifications, and each has an influence upon the size of material and the relative proportions of sand silt and clay.

Although tills are variable, some general relations are predictable within a local area. Clast percentage changes with transport distance. Clast size reduction depends upon the mode of transport: superglacial, englacial, or basal (Dreimanis, 1976). Tills from shale and limestone are primarily clay and silt. Tills from igneous and metamorphic rocks are mostly sand (Dreimanis & Vagners, 1972).

The upstream bedrock controls till composition in areas of shallow till, such as New England. In North Dakota, the coarsest tills are in areas where the

TABLE 3.2. Classification of Till Based Upon Process of Debris Release and Position Relative to the Glacier

Position	Primary	Secondary	
Subglacial	Meltout, sublimation, lodgment.	Deformation.	Settling through standing water.
Supraglacial	Meltout, sublimation.	Sediment flow, gravitational slumping.	

From Chorley et al., 1985, Table 17.4, p. 450; Modified from Lawson, 1981a. Reprinted by permission from S.A. Schumm.

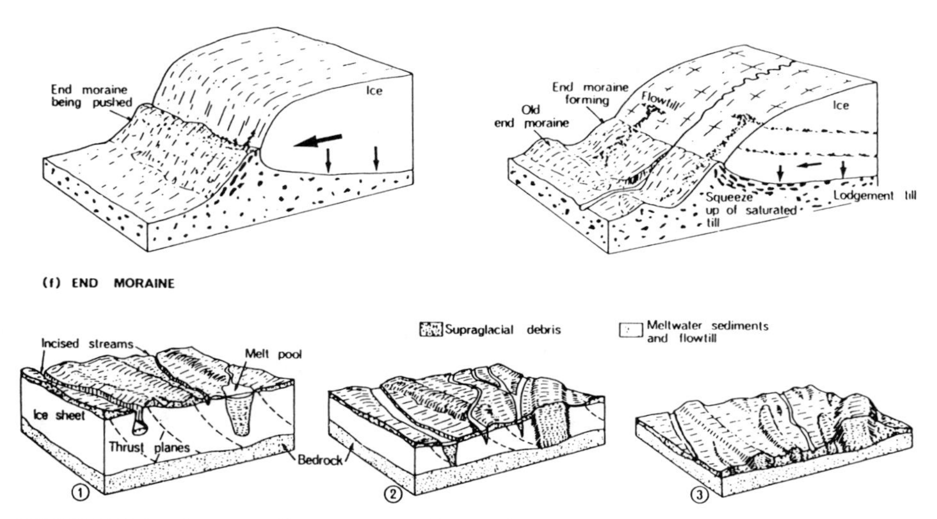

FIGURE 3.2. Cartoon of processes and deposits at a glacier front. From Selby, 1985, Fig. 15.19, p. 453. Reprinted by permission from M.J. Selby.

glacier moved over sand and gravel. The finest textured tills are where the glacier moved over large areas of lake sediment or Cretaceous shale, and till characteristics and mineralogy change appreciably over short distances (Clayton et al., 1980).

Depositional Environments Recent workers have subdivided glacial materials into a wide range of depositional environments based upon where the glacier carries the material within the ice and the depositional process. Clayton et al. (1980), Kemmis et al. (1981), and Ritter (1986) give good examples of the complexity of environments within till sheets.

A two-way subdivision probably is the most meaningful classification to field soil scientists. Till emplaced beneath the ice is lodgment or basal till. Ablation till is material carried within or upon the ice and is reworked by water to varying degrees before and during deposition. Lodgment till is fine-textured and usually has a high density compared to ablation till (Fig. 3.3). Lodgment till properties vary less than ablation till. Ablation till has several depositional environments and usually is coarser than associated lodgment or basal till (Figs. 3.4 and 3.5; Table 3.3).

The North Dakota Geologic map delineates several landforms closely associated with the depositional environment. Collapsed glacial sediment (hummocky topography), collapsed-draped transition, draped topography (slight modification of earlier topography), thrust masses, subglacially molded surfaces and eroded glacial sediments are the major landforms (Clayton et al., 1980, p. 34–52). Clayton and associates mapped several subdivisions of the above topography. The source determines till properties, and several useful

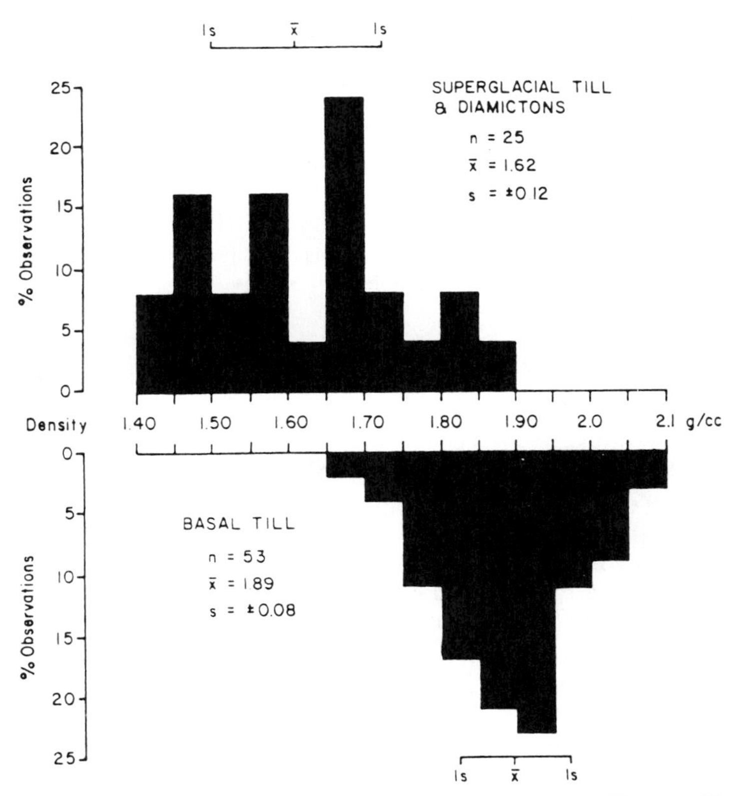

FIGURE 3.3. Bulk density of superglacial sediments and basal till. From Kemmis et al., 1981, Fig. 12, p. 34. Reprinted by permission from T.J. Kemmis, Iowa Geological Survey.

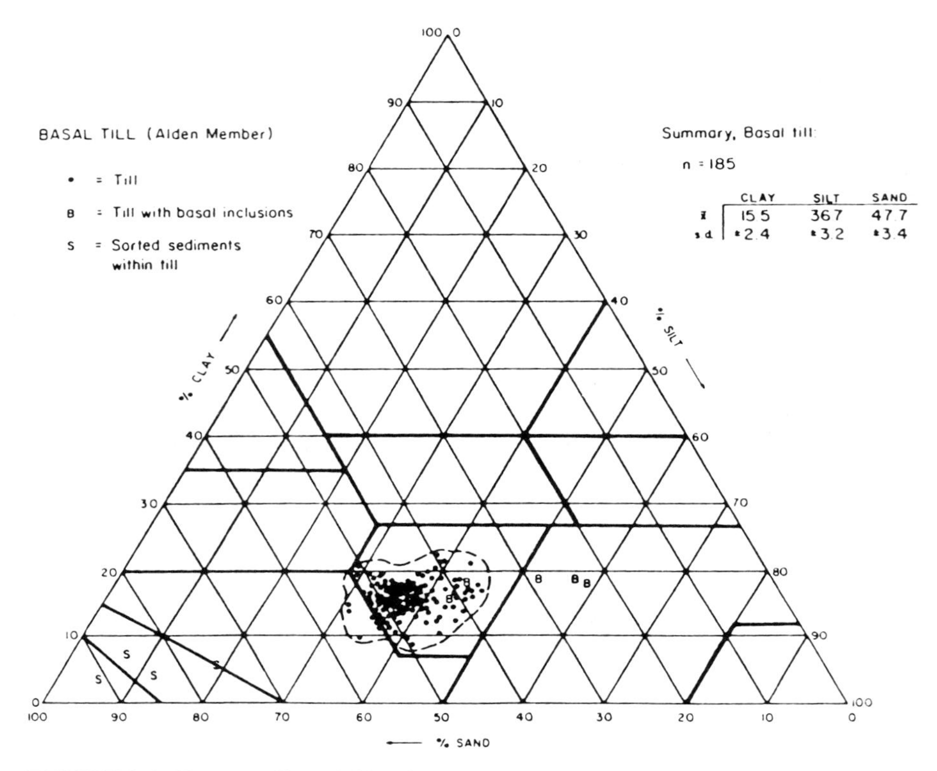

FIGURE 3.4. Texture of basal till. After Kemmis et al., 1981, Fig. 7, p. 27. Reprinted by permission from T.J. Kemmis, Iowa Geological Survey.

subdivisions based upon chemical and mineralogical compositions exist. For example, in North Dakota the tills from Pierre Shale are fine textured and may have large quantities of salt (Miller et al., 1981). In Iowa the older tills commonly are clay loam, but the younger Des Moines Lobe is mainly a calcareous loam (Kemmis et al., 1981).

Environments and Landforms The successful use of landforms in soils work requires that some relationship exist between surface topography and the depositional environment. Data by Kemmis et al. (1981) show a simple to complex relationship between landforms and glacial materials (Table 3.3; Fig. 3.6). Figure 3.7 is an example of the complex collapsed till landscape in North Dakota (Clayton et al., 1980). The soil and landform associations shown in Figures 3.6 and 3.7 demonstrate the complexity of the depositional processes.

The properties of soil materials can change abruptly in a glacial landscape and most soil maps are complexes of several soil series. Yet order exists within these complex areas. Basal till is close to the surface of the minor moraines of Gwynne (1942), Kemmis et al. (1981), and the transversely elongated hum-

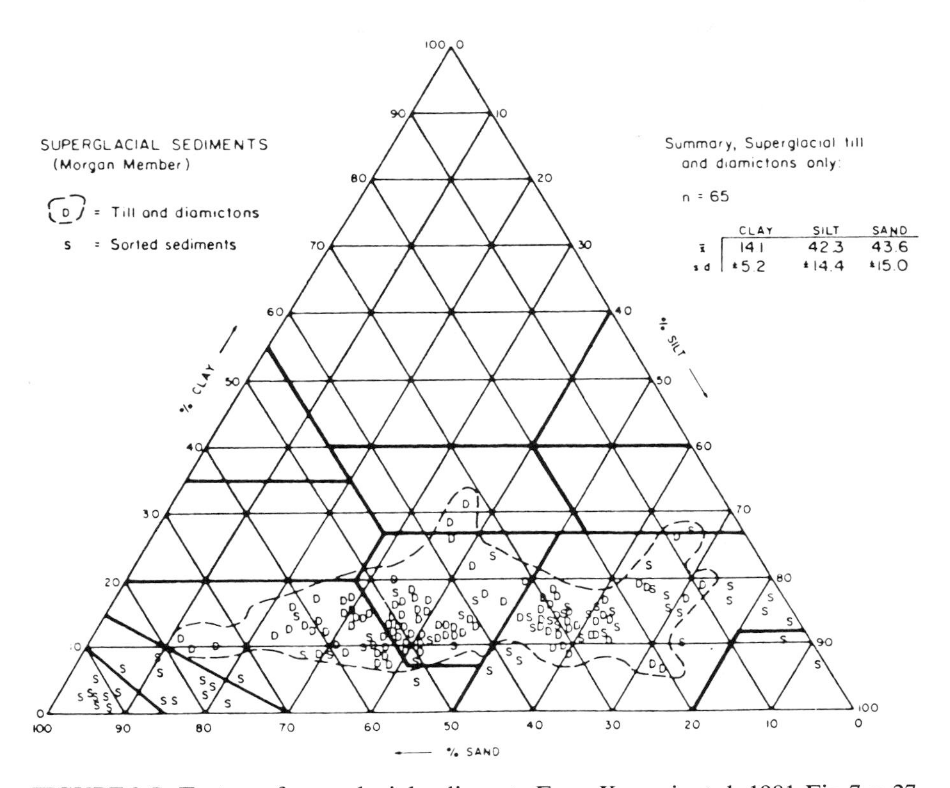

FIGURE 3.5. Texture of superglacial sediments. From Kemmis et al., 1981, Fig. 7, p. 27. Reprinted by permission from T.J. Kemmis, Iowa Geological Survey.

mocks of collapsed glacial sediment of Clayton et al. (1980). Landforms provide the field investigator with a useful predictive tool. We should predict both texture and bulk densities of the soil materials with reasonable accuracy by using the glacial landform.

In other glacial landscapes the details may change, but a relationship should exist between landforms and depositional environments. In parts of New York, the till bedrock source and distance of travel control soil properties (Cline, 1949). Soils near the outcrop of limestone have above normal pH and very high carbonate contents. A sequence of soils sampled downstream from the limestone outcrop shows a continuous predictable series of changes in physical and chemical properties. Sequential changes in soil properties are rare in areas of thick till and long-distance transport.

Table 3.3 can be a useful guide in predicting soil materials if there has been little post-depositional modification by erosional processes. Surface modification of young till sheets, such as the Des Moines Lobe in central Iowa, is the rule not the exception. Soils on the Des Moines Lobe are formed entirely or partly in Holocene materials eroded and redeposited by local erosional proc-

TABLE 3.3. Glacial Landforms and Depositional Environments

Outwash Valley Trains and Lake Basins	Depositional Environment
Outwash terraces	Concentrated in present river channels along lateral moraines.
Lake basinis	Concentrated near hummocky, high-relief areas near lateral moraines.
Upland Glacial Landforms	
Prominent ridges, high to moderate local relief, irregular hummocky topography	Ice-marginal (morainal) High-relief hummocks and closed depressions with 10–20 m thick supraglacial sediments.
Irregular hummocky topography on broad topographic swells and swales of moderate relief	Moderate relief hummocks (3–10 m) on gentle swells and swales. Composed of 3–6 m thick supraglacial sediments over basal till.
Irregular hummocky topography with common circular depressions on broad swells and swales of *moderate relief.*	Three to six m thick supraglacial sediments over basal till.
Irregular hummocky topography with common circular depressions on broad swells and swales of *moderate to high relief.*	Three to six m thick supraglacial sediments over basal till. Extensive areas of high relief.
Irregular hummocky topography with *circular ridges and depressions* inset upon broader swells and swales with *moderate to high relief.*	Three to six m thick supraglacial sediments over basal till. Both ridges and depressions are
Small curvilinear ridges, 1–2 m relief on broad swells and swales.	The minor moraines and swell and swale topography of others. Spacing between ridges averages 105 m, or about 9 ridges/km. Ridges are discontinuous but show prominently on aerial photos. Ridges are composed of thick, uniform, relatively dense basal till overlain by thin slopewash and colluvial sediments.
Discontinuous curvilinear ridges.	Three to 6 m supraglacial sediments over moderately thick basal till.
Upland plain with diffuse ridges and depressions.	A veneer (0.5–2 m) cover of fine-textured glaciolacustrine sediments over till-like sediments.
Gently sloping plain with small circular depressions.	Thin fine-textured glaciolacustrine sediments over till-like sediments.
Plain, lower than adjacent areas, with few large irregular depressions.	Complex sequence of supraglacial till-like sediments and meltwater deposits plus thick outwash sands and gravel.

From Kemmis et al., 1981, Fig. 2, p. 5. Reprinted by permission from T.J. Kemmis, Iowa Geological Survey.

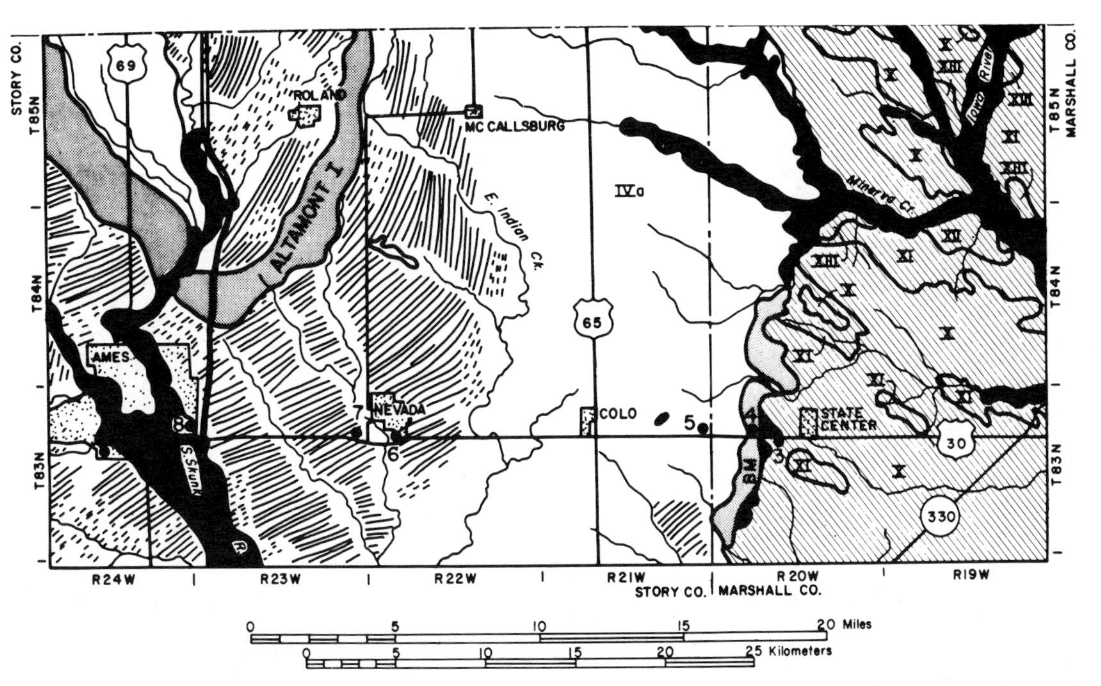

FIGURE 3.6. Relationship between landform and depositional environment in part of the Des Moines Lobe. *Late Wisconsinan & Holocene*: Black = outwash terraces and recent flood plains. *Late Wisconsinan*: Alamont I and BM = moraines with 5–20 m supraglacial sediment. IVa = irregular hummocky topography, 3–6 m supraglacial sediment over basal till. Parallel curved lines = curvilinear ridges, 1–2 m relief dominated by basal till > 5 m thick. Short dashed lines = discontinuous or poorly expressed ridge forms dominated by basal till with common supraglacial sediment 1–5 m thick. *Wisconsinan or older*: Hachured area = stream dissected landscapes with loess, buried soils and eolian sands. From Kemmis et al., 1981, Fig. 2, p. 6. Reprinted by permission from T.J. Kemmis, Iowa Geological Survey.

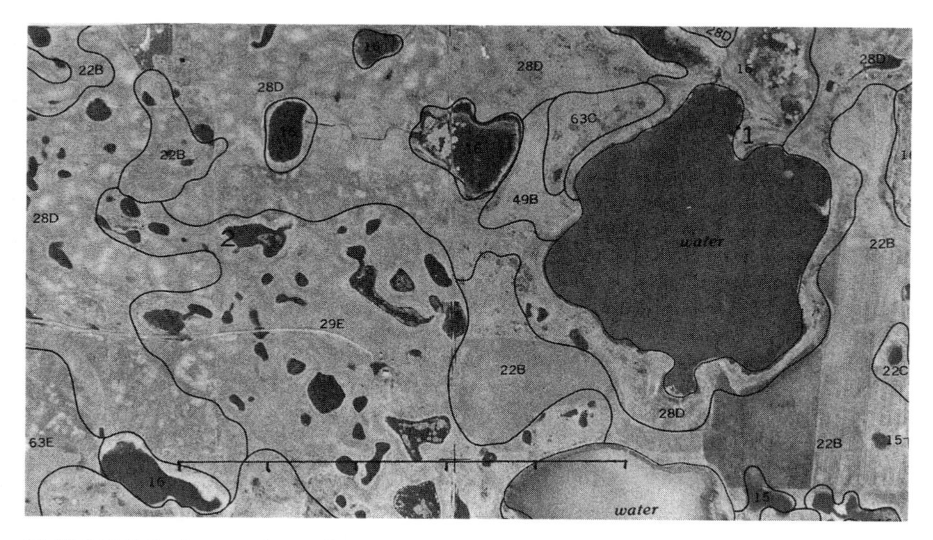

FIGURE 3.7. A complex soil-material landscape of collapsed till in North Dakota. *Legend*: 15 = Parnell Sicl. 16 = Southam Sicl. 22B = Barnes-Seva Loams 1–6% slope. 22C = Barnes-Buse Loams 6–9% slope. 29E = Barnes-Buse-Parnell Complex 0–35% slope. 49B = Arvilla Sandy Loam 1–6% slope. 63C = Sioux-Arvilla Sandy Loams 1–9% slope. 63E = Sioux-Arvilla Sandy Loams 9–35% slope. From Seelig and Gulsvig 1988, northeast corner of map sheet 4.

esses (Burras & Scholtes, 1987; Walker, 1966). Where the reworked materials are thicker than the solum, the soils may have little relation to the depositional environment of the drift shown on detailed geologic maps.

Eolian Deposits

Eolian deposits are widespread and important soil materials that can occur wherever wind moves over barren sand, silt or clay. The two chief eolian materials that occupy large areas of the world are sands and loess (Chorley et al., 1985; Cooke & Warren, 1973; Smalley & Vita-Finzi, 1968; Snead, 1972). Sands are traction deposits, and loess is a suspension deposit. Eolian clay (parna) is a minor but locally important soil material that is both a traction and suspension deposit. Aerosols, very fine-grained materials carried in suspension to very high altitudes and over large distances, are a fourth eolian material. Few, if any, pure aerosol deposits are thick enough to contain the soil solum.

Eolian materials can add considerable complexity to the landscape. They can be discontinuous deposits that collect in the lee of plants, coppice dunes, or deposits that mantle the landscape. Dune sands with a swell and swale topography hide or subdue the contours of the underlying surfaces. The large sand seas (McKee, 1979) usually have a complex of dune and inter-dune deposits with considerable variation in surface form and textures (Ahlbrandt, 1979).

Uneroded loess deposits usually have uniform local thickness and mimic the contours of the buried surface (Daniels & Young, 1968). Eroded thin loess deposits on irregular relief produce a heterogeneous soil landscape (Robertus et al, 1989). Areas of dissected dunes and loess deposits have complex material outcrop patterns from the irregular exposure of buried soils and underlying sediments.

Eolian Sands Eolian sands are common in many environments (McKee, 1979). They occur at seashores, along river valleys, in deserts, and at the edges of lakes or other water bodies. Any process that exposes sands to wind can result in a downwind deposit of eolian sand. Dunes from blowouts are common in sandy terrains. In humid regions, many eolian sand deposits are relict features of an earlier period of landscape and climatic instability. The Nebraska sand hills area is an example of a large body of relict dunes.

Textural properties. Ahlbrandt (1979) divided eolian sands into coastal, inland and interdune environments. Coastal dune sands are largely fine sands and are well sorted to very well sorted (Ahlbrandt, 1979). Moderate to well-sorted fine to medium sands dominate inland dunes (Fig. 3.8). The associated interdune sands have bimodal sand fractions and a higher content of silt and clay than the adjacent dune-sand samples. The coastal and inland dune sands range in mean grain size from 1.6 to 0.1 mm. The position of the sample within the dune and the type of dune also influence the grain size (Fig. 3.9). Most eolian sands have some grains >2 mm and a low, <10%, silt and clay content.

Poorly sorted fluvial deposits are the source for many dunes. The wind winnows the fines and leaves behind most of the coarse particles (Fig. 3.10). Saltation populations are dominant, but absolute grain size depends upon wind velocity and sand source. Grains become airborne by aerodynamic lift or from impact of other saltating grains returning to the surface. The saltating grains generally rise less than a meter from the surface before descending.

Eolian sand grains have a surface grain morphology of sharp edges, conchoidal surfaces and striations. Water-transported and sorted sands have rounded and polished surfaces. Thus, it is possible to infer depositional environment from grain morphology. However, sand grains can retain surface characteristics obtained in another environment, and interpretation requires caution.

About four-fifths of the sand moved by wind is by saltation (Bagnold, 1941). Grains too large to saltate move by surface creep from wind shear on the bed and by the impact of smaller saltating grains. Clay adhering to the sand grains is the source of much of the clay in eolian sand bodies. Later additions of fines are possible from aerosols and from weathering of primary minerals (Walker, 1979). Table 3.4 gives the time of flight and transport distance of particles from

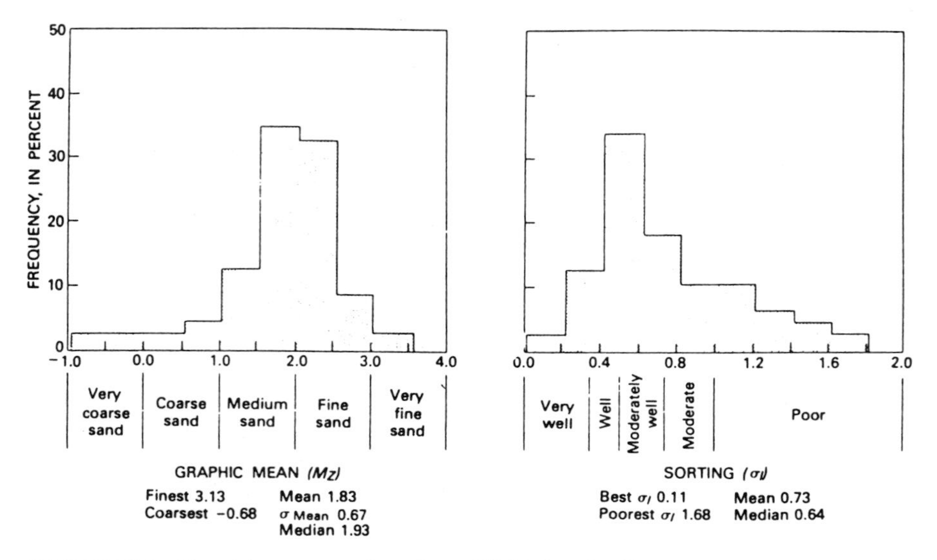

FIGURE 3.8. Graphic mean and sorting of 191 inland eolian sand samples. Modified from Ahlbrandt, 1979, Fig. 18, p. 28.

medium sand to clay in a 15 meters/second (m/s) wind (Chorley et al., 1985). Pethick (1984) has an excellent discussion of the processes involved in coastal dune formation, and we recommend his work for a detailed discussion of eolian sand movement and capture.

Distribution and Form. Eolian sands can move several kilometers from their source and develop large sand seas in areas of abundant source and sparse vegetation (Breed & Grow, 1979; Breed et al., 1979; Gile, 1979). Dune sands in humid areas commonly occur as narrow bands on the leeward sides of valley reaches (Fig. 3.11) or next to open bodies of water. Deposition of sand in dense vegetation, weathering and soil formation destroys the bedding planes in many of these sand bodies. Table 3.5 outlines the major dune forms. For additional classification of sand dunes see Thomas (1989, Fig. 11.6, p. 242).

Dune sands have a narrow textural range (Figs. 3.8 and 3.9), but soils formed in these sands have a variety of morphologic features depending upon the moisture at each site. In semiarid regions, episodic deposition of sand results in a complex of deposits with similar textures, but different soil morphology and age (Gile, 1979). Most eolian sand bodies in humid areas are complexes of several recognized soil series. The ages may be similar, but the rapid changes in topography and moisture regimes have corresponding changes in soil morphology and plant growth potential. The location of specific soils within a dune landscape is predictable.

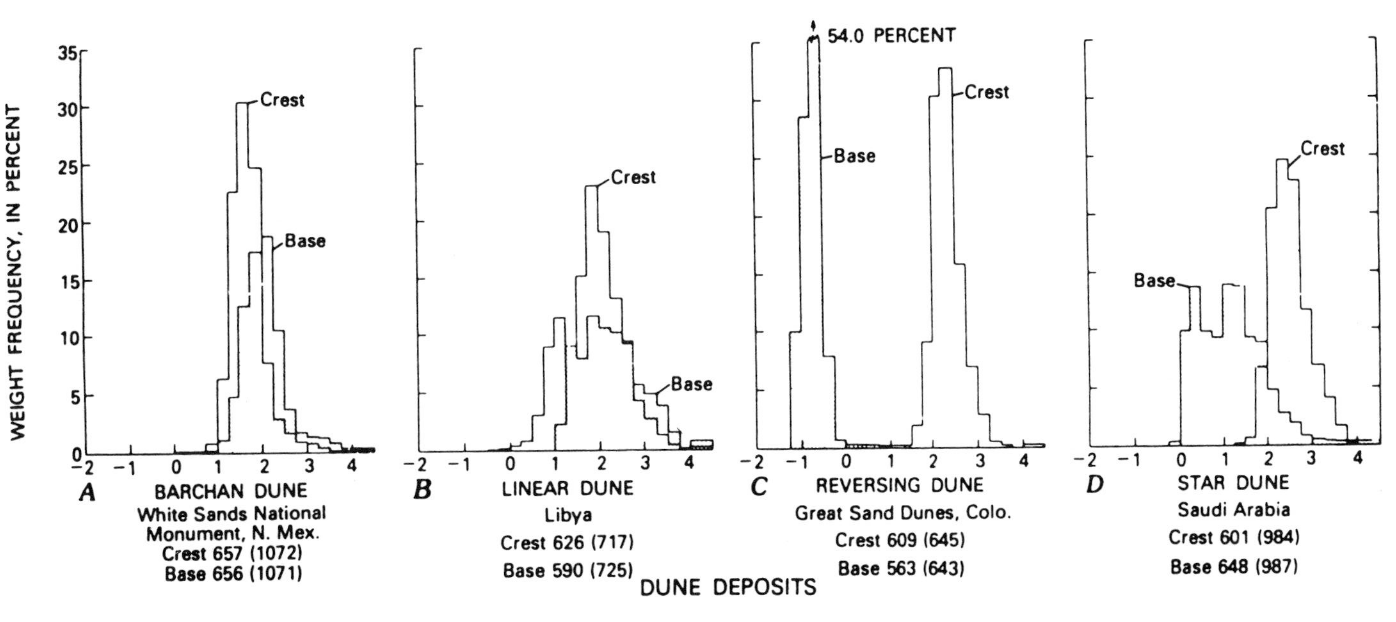

FIGURE 3.9. Textural contrasts within a dune and between dune types. From Ahlbrandt, 1979, Fig. 22 A–D, p. 31.

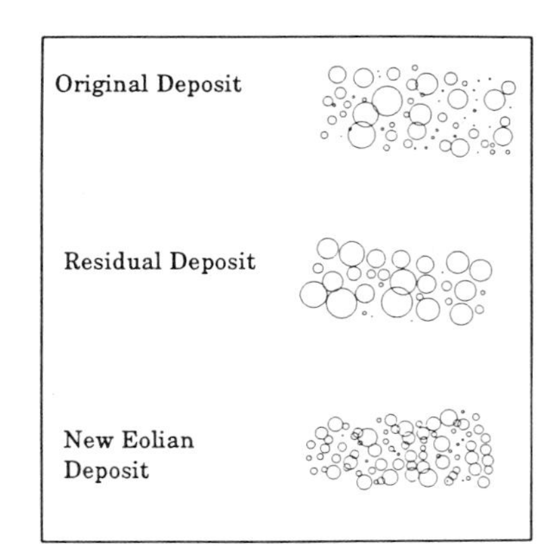

FIGURE 3.10. Selective removal of finer material by wind. Redrawn from Folk, 1968, Fig. 7, p. 23.

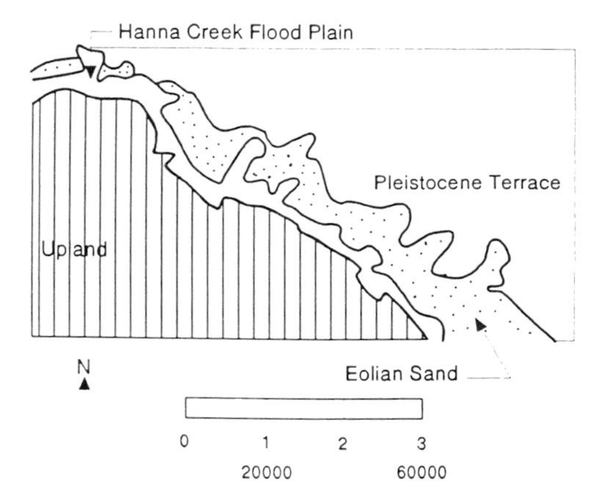

FIGURE 3.11. Relation of eolian sands to stream valley orientation, Hanna Creek Area, Johnston County, North Carolina. Redrawn from Johnston County, N.C. unpublished Soil Survey Sheet, Soil Conservation Service, USDA.

TABLE 3.4. Time of Flight and Transport Distance for Particles in a 15 m/s Wind

Size	Flight Time	Transportation Distance (km)	Moved by Saltation (%)
Clay	9–90 yrs.	$4–40 \times 10^6$	0
Silt	8–80 yrs.	$4–40 \times 10^2$	0
Very fine sand	0.3–3 s	46–460 m	?
Fine sand	very variable		84
Medium-coarse sand	very variable		75

Modified from Chorley et al., 1985, Table 16.3, p. 427. Reprinted by permission from S.A. Schumm.

TABLE 3.5. Basic Dune Types

Name	Form	Comments
Sheet	Sheet-like with broad, flat surface, no dune forms.	
Stringer	Thin, elongate strip on bedrock.	
Dome	Circular or elliptical mound.	Modified barchans?
Barchan	Crescent in plan view.	Small sand supply, one dominant wind direction.
Barchanoid	Row of connected crescents.	Moderate sand supply, one dominant wind direction.
Transverse	Asymmetrical ridge.	Large sand supply, one dominant wind direction.
Blowout	Circular rim of depression.	Deflation, vegetation control.
Parabolic	U-shape.	As above.
Linear	Symmetrical ridge.	Bimodal dominant wind direction.
Reversing	Asymmetrical ridge.	Intermediate between star dune and transverse ridge.
Star	Central peak with 3 or more arms.	Tend to grow vertically.

From Chorley et al., Table 16.1, p. 416–417. Originally from McKee, 1979 and Breed and Grow, 1979. Reprinted by permission from S.A. Schumm.

Loess Loess is an eolian deposit dominated by silt and covers about 10% of the land surface (Selby, 1985). It has a regional association with a source area, but unlike eolian sands in humid regions, it usually mantles a much larger area. Most loess sheets are in or near glaciated regions. No mechanism other than glacial grinding can produce appreciable quantities of silt-sized quartz particles (Smalley, 1971, 1975).

Loess is a silty material of eolian origin. However, silty deposits can form in many ways, and not all silt deposits are loess. Loess eroded and redeposited by fluvial processes is a new alluvial deposit, and the new soil parent material is alluvium. A loess-mantled landscape that has been dissected can have a mosaic of loess, alluvium and exhumed paleosols. Time zero for soil development in the alluvium and truncated loess slopes is the end of deposition of the alluvium to which the truncated slopes grade. To understand the areal distribution and properties of the landscape requires the recognition of the mode of transport and deposition of soil materials. The terms "reworked" or "secondary loess" are misleading—and result in faulty interpretation of soil profile development.

Texture. Unweathered thick deposits of loess usually are texturally uniform vertically and horizontally within a local area. Loess thickness and texture change progressively with distance from the source (Fig. 1.2A; Figs. 3.12A, B and C; Smith, 1942). The thick Midwest loess deposits have low sand contents,

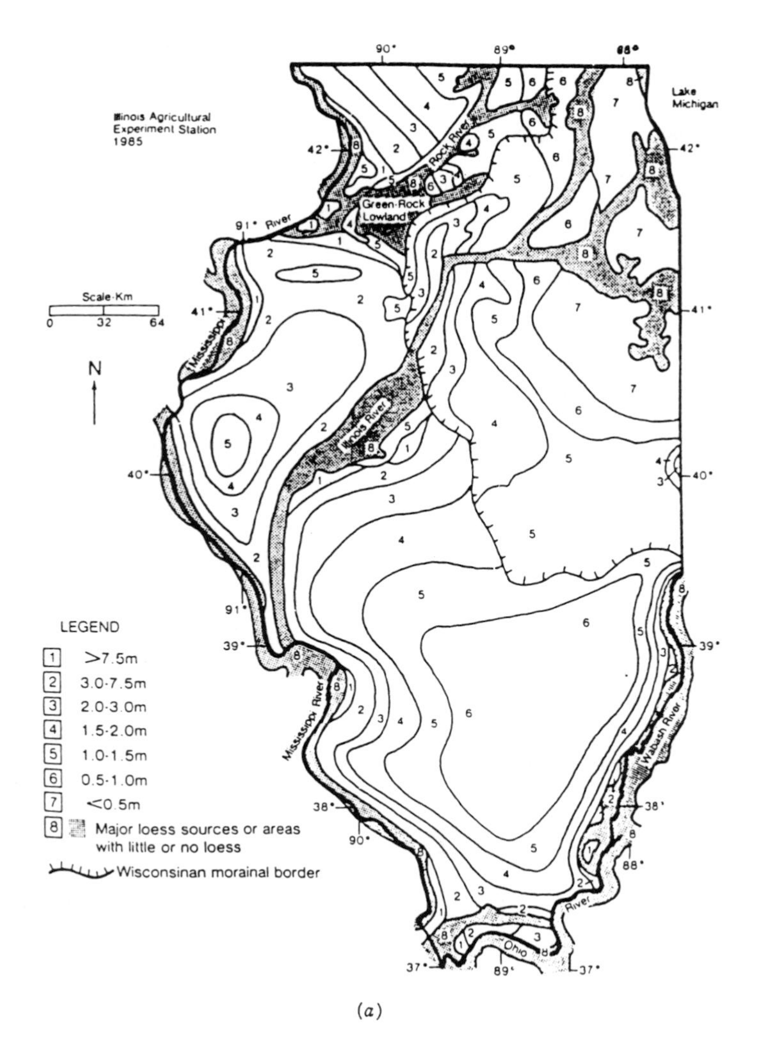
Illinois Agricultural
Experiment Station
1985
Lake
Michigan
Rock River
Green-Rock
Lowland
River
Mississippi
Illinois River
Mississippi River
Wabash River
Ohio
River
Scale-Km
0
32
64
N
LEGEND
1 >7.5m
2 3.0-7.5m
3 2.0-3.0m
4 1.5-2.0m
5 1.0-1.5m
6 0.5-1.0m
7 <0.5m
8 Major loess sources or areas with little or no loess
Wisconsinan morainal border

(*a*)

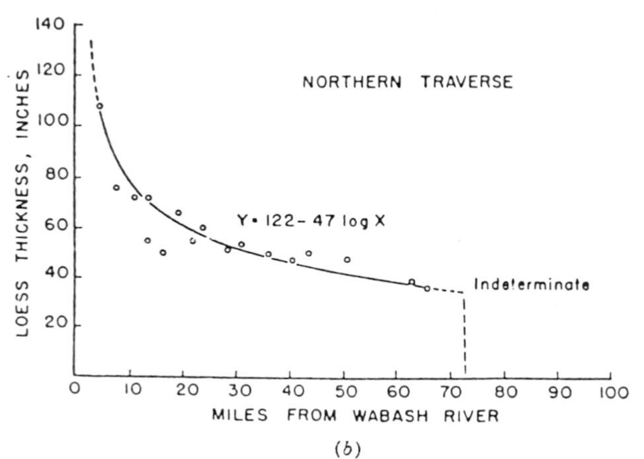
NORTHERN TRAVERSE
Y = 122 - 47 log X
Indeterminate
LOESS THICKNESS, INCHES
MILES FROM WABASH RIVER
140
120
100
80
60
40
20
0
10
20
30
40
50
60
70
80
90
100

(*b*)

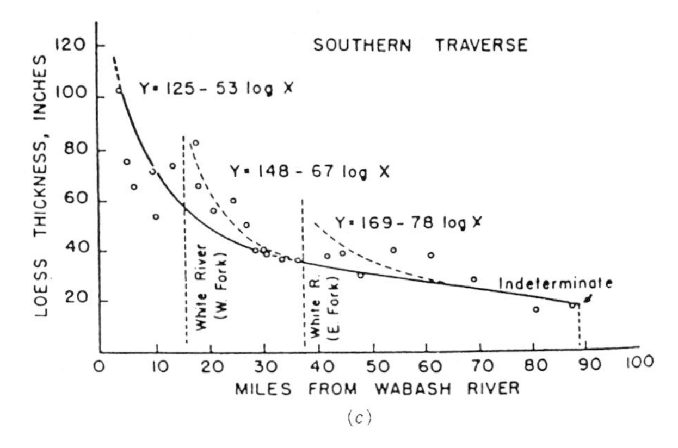

FIGURE 3.12. Regional changes in loess with distance from source. *A.* In Illinois. From Fehrenbacher et al., 1986, Fig. 1, p. 425. *B.* In Indiana (northern traverse). From Caldwell & White, 1956, Fig. 2, p. 261. *C.* In Indiana. From Caldwell & White, 1956, Fig. 3, p. 261. A is reprinted by permission from Williams and Wilkins Publishers, J.B. Fehrenbacher and K.D. Olson. B and C are reprinted by permission from J.L. White.

commonly less than 2%, and most of the sand is between .05 and .10 mm. Silt is the dominant size fraction in the unmodified deposit, and the clay content varies with the weathering environment and distance from source. Clay can dominate in thin loess deposits, at least in the soil B horizon.

Not all loess deposits are dominantly silt. The Palouse loess in eastern Washington has a very high sand content and is interbedded with sand and tephra (Busacca, 1989, Fig. 2, p. 110). The Palouse loess has a much more complex depositional environment than the thick deposits associated with valley outwash in the Midwest.

Loess usually is high in weatherable minerals, and the silt and clay mineralogy depends upon the source (Kleiss & Fehrenbacher, 1973). Feldspar content, calcium, sodium and potassium change with distance from source either from weathering or from changes related to particle-size changes away from the source (Caldwell & White, 1956; Ruhe et al., 1967; Smith, 1942). Loess can be calcareous or noncalcareous, although some authors define it as only being calcareous.

Other Properties. Vertical-cut banks in thick loess deposits are stable if dry (Lohnes & Handy, 1968). Water flowing across the face or seeping from the toe quickly erodes loess banks. These deposits usually have little visible bedding and are difficult to separate from adjacent brown or tan alluvial deposits.

Loess deposits thin slowly with distance from source if the winds were at right angles to the river valley source. Thinning from the source valley is rapid if the winds are parallel to the river valley (Handy, 1976). Loess in Iowa consolidates, collapses about 5% when first saturated (Handy, 1973), and can cause considerable structural damage during the initial saturation. Wang et al. (1987), Lautridou et al. (1987), and Lutenegger (1987) discuss the geotechnics of loess in various parts of the world. Wen et al. (1987) detail the geochemical environment of loess in a section of China. The entire *Catena* supplement edited by Pecsi (1987) is a good review of some properties of loess.

Local Thinning. Loess mantles the landscape and the surface of an uneroded loess deposit mimics the surface of the underlying material (Daniels & Young, 1968). Some of the variation in loess thickness in a local area correlates with the elevation of the buried surface. The deposit thins from primary to secondary, and tertiary divides both toward and away from the source (Ruhe, 1954).

Deposition of loess sheets requires several hundreds of years, and some modification of the sloping areas is probable. The thinning of loess on valley slopes and the exposure of buried soils and sediments are largely from post-depositional erosion. The evidence for this is the truncation of weathering zones by valley slopes in thick loess areas (Fig. 1.3A; Ruhe et al., 1955).

Soils. Soil differences within a local uneroded landscape are the effect of the spatial and temporal distribution of water within that landscape. Topographic factors including position within the landscape, surface shape, aspect, and surface stability will control differences in depth to the clay or salt maximum and its thickness. The amount of illuvial material or clay within the zone of maximum accumulation is relatively constant in Pleistocene and Holocene landscapes. The nearly level, stable surfaces usually have the thickest and deepest zone of maximum clay accumulation (Ruhe et al., 1967).

Truncation of slopes by Holocene erosion increases the variability of soil materials. Truncation exposes different weathering and mineralogic zones and results in local sorting and deposition of materials on the slopes. Later additions of loess near major river valleys can result in a very complex loess stratigraphy.

Loess probably is the ideal material to use when studying soil genesis, but one must know the local geomorphology. One should work with one stratigraphic unit of similar mineralogy that has a narrow duration of weathering. For example, in southwest Iowa the clay content of the clay maximum is similar for most loess-derived soils within a local landscape (Ruhe et al., 1967). The depth to the clay maximum varies with geomorphic surface or duration of weathering.

Smeck and Runge (1971) reported major morphological and chemical differences in soils on a broad loess-mantled divide in Illinois. The soils are all the same age and developed in the same part of the loess (assuming 0 erosion). Soil differences are the result of different hydrologic environments, not material or time.

In western Iowa loess, erosion exposed calcareous brown and calcareous gray zones the valley slopes (Ruhe et al., 1955). Texturally the soils are similar, but the materials did not have the same properties when initially exposed to soil formation. It is as difficult in loess as it is in other materials to hold all soil-forming factors constant but the one of interest.

Parna Parna is an eolian clay deposited as sand and gravel size clay aggregates. Butler (1956) coined the term *parna*, but Coffey (1909) first described eolian clay dunes. The clay aggregates contain accompanying sand and silt grains bonded in part by saline evaporite crystals. In south Texas the clay aggregates develop dunes (Huffman & Price, 1949), but in western Australia they also are sheet deposits downwind from dunes (Fig. 3.13; Bettenay, 1962).

The parna deposits in Australia studied by Butler and Hutton (1956) increase in clay content downwind from the associated sand dune crest whereas nonclay sizes decrease. Parna or eolian clay deposits occur in a semiarid climate that has intermittent saline lakes, lagoons or saline mud flats (Butler, 1956; Huffman & Price, 1949). The clayey surfaces crack when dry, and the wind then transports the clay aggregates and associated sand and silt

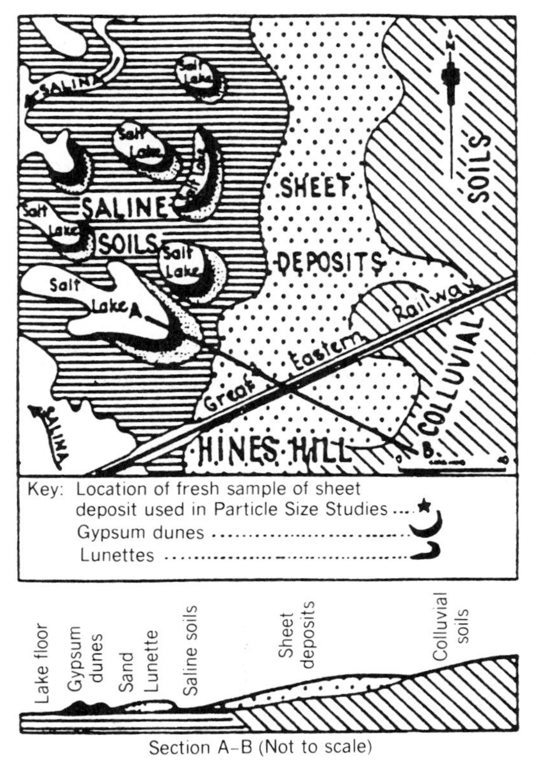

FIGURE 3.13. Eolian deposits associated with saline lakes in western Australia. From Bettenay, 1962, Fig. 2, p. 12. Reprinted by permission from Blackwell Scientific Publications.

grains by saltation or creep. In Australia the aggregates are difficult to disperse. As with other eolian materials, parna also has predictable changes in properties and association with other landscape features and environments.

Aerosols Aerosols are 0.002 to 0.010 mm material transported from deserts and other sources (Syers et al., 1969). This fine atmospheric dust can travel around the world for several months or years before being deposited (Table 3.4). Estimates of atmospheric dust deposition on Barbados are 0.6 meter each 1,000 years (Delaney et al., 1967), and in permanent snow fields 1 to 10 meters each million years (Windóm, 1969).

Middleton (1989, Table 12.8) gave a range of dust deposition ranging from 14 to 200 t/km^2/yr or tons per square kilometer per year for several arid areas. Amundson et al. (1989) suggested the possibility of 6 cm of dust added to the soil in the last 10,000 years; they concluded that dust has a major role in formation of the platy, vesicular surface soil horizon. Wilding et al. (1989) measured a dustfall of about 1500 kg/ha/yr between 1985 and 1987 in Niger, Africa. Measurements were during the harmattan for intervals of two to six months. The mean clay content was 14%, although the range was 0.1 to 27.5. The dust samples had 10 to 30 times more bases than the reference soils. For more information on the physical, chemical and mineralogical composition of African dust see Danin and Ganor (1991).

Inoue and Naruse (1987) estimated that eolian dust from China deposited 4 to 23 mm/1000 years in Japan. Mizota et al. (1991) measured up to 2 m of a silty mantle in Japan, more than 15° latitude from the assumed source. The oxygen isotope composition of quartz in the mantle overlying quartz-free volcanic material was very similar to the ratios from loess in China. Also, the particle-size distribution of the quartz decreased from west to east, a feature supporting the eolian deposition of the mantle.

There is little question that aerosols are ubiquitous materials, but pure surface deposits of measurable thickness are unknown. Delany and Windom's previously cited estimates would place 1 to 20 meter of aerosols on a stable end tertiary coastal plain surface. The aerosol minerals weather easily in humid climates, their depositional rates are low and mixing by fauna and flora prevents their recognition as separate deposits. But these materials can influence the mineralogical properties of soils and sediments. What influence do aerosols have on surface and subsurface clay mineralogy of soils on ancient stable geomorphic surfaces?

ADDITIONAL READING

Edwards, M. (1986). Glacial environments. In *Sedimentary Environments and facies* (pp. 445–470). Ed. by H.G. Reading. Boston: Blackwell.

Collison, J.D. (1986). Deserts. In *Sedimentary Environments and facies* (pp. 95–112). Ed. by H.G. Reading. Boston: Blackwell.

REFERENCES

Ahlbrandt, T.S. (1979). Textural parameters of eolian deposits. In *A Study of Global Sand Seas* (pp. 21–51). Ed. by E.D. McKee. U.S. Geol. Survey Prof. Paper, No. 1052.

Amundson, R.G., O.A. Chadwick, J.M. Sowers, and H.E. Doner. (1989). *Geoderma*, 43:349–371.

Bagnold, R.A. (1941). *The Physics of Blown Sand and Desert Dunes*. London: Methuen.

Bettenay, E. (1962). *J. Soil Sci.*, 13:10–17.

Breed, C.S. and T. Grow. (1979). Morphology and distribution of dunes in sand seas observed by remote sensing. In *A Study of Global Sand Seas* (pp. 253–302). Ed. by E.D. McKee. U.S. Geol. Survey Prof. Paper, No. 1052.

Breed, C.S. et al. (1979). Regional studies of sand seas, using Landsat (ERTS) imagery. In *A Study of Global Sand Seas* (pp. 305–397). Ed. by E.D. McKee. U.S. Geol. Survey Prof. Paper, No. 1052.

Burras, C.L. and W.H. Scholtes. (1987). *Soil Sci. Soc. Amer. J.*, 51:1541–1547.

Busacca, A.J. (1989). *Geoderma*, 45:105–122.

Butler, B.E. (1956). *Aust. J. Sci.*, 18:145–151.

Butler, B.E. and J.T. Hutton. (1956). *Aust. J. Agric. Res.*, 7:536–555.

Caldwell, R.E. and J.L. White. (1956). *Soil Sci. Soc. Amer. Proc.*, 20:258–263.

Clayton, L., S.R. Moran, and J.P. Bluemle. (1980). *Report of Investigations, No. 69*. North Dakota Geological Survey.

Cline, M.G. (1949). *Soil Sci.*, 68:259–272.

Cline, M.G. (1963). *Soil Sci.*, 96:17–22.

Chorley, R.J., S.A. Schumm, and D.E. Sugden. (1985). *Geomorphology*. New York: Methuen.

Coffey, G.N. (1909). *J. Geol.*, 17:754–755.

Cooke, R.U. and A. Warren. (1973). *Geomorphology in Deserts*. London: Batsford.

Daniels, R.B. and K.K. Young. (1968). *SE Geol.*, 9:9–19.

Danin, A. and E. Ganor. (1991). *Earth Surface Processes and Landforms*, 16:153–162.

Delaney, A.C. et al. (1967). *Geochemica et Cosmochemica Acta*, 31:885–929.

Dreimanis, A. (1976). Tills: Their origin and properties. In *Glacial Till* (pp. 11–49). Ed. by R.F. Legget. Royal Society of Canada, Spec. Pub. 12.

Dreimanis, A. and U.J. Vagners. (1972). Bimodal distribution of rock and mineral fragments in basal till. In *Till: A Symposium* (pp. 237–250). Ed. by R.P. Goldthwait. Columbus: Ohio State Univ. Press.

Fehrenbacher, J.B., K.R. Olson, and I.J. Jansen. (1986). *Soil Sci.*, 141:423–431.

Folk, R.L. (1968). *Proc. Int. 23rd Geol. Congress*, Sec. 8, (pp. 9–32). Ustredni Ustav Geologicky, Praha Prague, Czechoslovakia.

Gile, L.H. (1979). *Soil Sci. Soc. Amer. J.*, 43:994–1003.

Goldthwait, R.P. (Ed.) (1972). *Till: A Symposium*. Columbus: Ohio State Univ. Press.

Gwynne, C.S. (1942). *J. Geol.*, 50:200–208.

Handy, R.L. (1973). *Soil Sci. Soc. Am. Proc.*, 37:281–284.

Handy, R.L. (1976). *Geol. Soc. Am. Bull.*, 87:915–927.

Huffman, G.G. and W.A. Price. (1949). *J. of Sed. Petrol.*, 19:118–127.

Inoue, K. and T. Naruse. (1987). *Soil Sci. Plant Nutr.*, 33:327–345.

Kemmis, T.J., G.R. Hallberg, and A.J. Lutenegger. (1981). *Depositional Environments of Glacial Sediments and Landforms on the Des Moines Lobe, Iowa*. Iowa Geological Survey, Guidebook Series, No. 6.

Kleiss, H.J. and J.B. Fehrenbacher. (1973). *Soil Sci. Soc. Am. Proc.*, 37:291–295.

Lautridou, J.P., M. Masson, and R. Voiment. (1987). Loess et geotechnique. In *Loess and Environment* (pp. 11–25). Ed. by M. Pecsi. Catena Supplement 9.

Lawson, D.E. (1981a). *CRREL Report*, 81–27.

Lawson, D.E. (1981b). *Annals of Glaciology*, 2:78–84.

Lewis, D.W. (1984). *Practical Sedimentology*. Stroudsburg, PA: Hutchinson Ross.

Lohnes, R.A. and R.L. Handy. (1968). *J. Geol.*, 76:247–258.

Lutenegger, A.J. (1987). In situ shear strength of friable loess. In *Loess and Environment* (pp. 27–34). Ed. by M. Pecsi. Catena Supplement 9.

Mizota, C., H. Endo, K.T. Um, M. Kusakabe, M. Noto, and Y. Matsuhisa. (1991). *Geoderma*, 49:153–164.

McKee, E.D. (Ed.) (1979). *A Study of Global Sand Seas*. U.S. Geological Survey Prof. Paper, No. 1052.

Middleton, N.J. (1989). Desert dust. In *Arid Zone Geomorphology* (pp. 262–283). Ed. by D.S.G. Thomas. New York: Halsted Press.

Miller, M.R. et al. (1981). *Agricultural Water Management*, 4:115–141.

Miner, G.S., S. Traore, and M.R. Tucker. (1986). *Agron. J.*, 78:291–295.

Pecsi, M. (Ed.) (1987). *Loess and Environment*. Catena Supplement 9.

Pethick, J. (1984). *An Introduction to Coastal Geomorphology*. Baltimore: Edward Arnold.

Ritter, D.F. (1986). *Process Geomorphology*. Dubuque, IA: Brown.

Robertus, R.A., J.A. Doolittle, and R.L. Hall. (1989). *Soil Sci. Soc. Amer. J.*, 53:843–847.

Ruhe, R.V. (1954). *Am. J. Sci.*, 252:663–672.

Ruhe, R.V., R.B. Daniels, and J.G. Cady. (1967). *Landscape Evolution and Soil Formation in Southwestern Iowa*. USDA Tech. Bull. 1349.

Ruhe, R.B., R.C. Prill, and F.F. Riecken. (1955). *Soil Sci. Soc. Amer. Proc.*, 19:345–347.

Seelig, B. and A.R. Gulsvig. (1988). *Soil Survey of Kidder County, North Dakota*. USDA Soil Conservation Service.

Selby, M.J. (1985). *Earth's Changing Surface*. Oxford: Clarendon.

Smalley, I.J. (1971). *Earth Sci. Rev.*, 7:67–85.

Smalley, I.J. (1975). *Loess Lithology and Genesis*. [Benchmark Papers in Geology, No. 20] New York: Wiley.

Smalley, I.J. and C. Vita-Finzi. (1968). *J. of Sed. Petrol.*, 38:766–744.

Smeck, N.E. and E.C.A. Runge. (1971). *Soil Sci. Soc. Amer. Proc.*, 35:952–959.

Smith, G.D. (1942). *Illinois Loess, a Pedologic Interpretation*. Ill. Agric. Exp. Sta. Bull., 490:139–184.

Smith, G.D. (1963). *Soil Sci.*, 96:6–16.

Snead, R.E. (1972). *Atlas of World Physical Features*. New York: Wiley.

Soil Survey Staff. (1975). *Soil Taxonomy*. [Agric. Handbook 436] Washington, DC: U.S. Govt. Printing Office.

Syers, J.K., M.L. Jackson, V.E. Berkheiser, R.N. Clayton, and R.W. Rex. (1969). *Soil Sci.*, 107:421–427.

Thomas, D.S.G. (1989). Aeolian sand deposits. In *Arid Zone Geomorphology* (pp. 232–261). Ed. by D.S.G. Thomas. New York: Halsted Press.

Walker, P.H. (1966). *Postglacial Environments in Relation to Landscape and Soils on the Cary Drift, Iowa*. Iowa State Univ. Ag. and Home Ec. Exp. Sta. Res. Bull., 549:838–875.

Walker, T.R. (1979). Red color in dune sand. In *A Study of Global Sand Seas* (pp. 61–81). Ed. by E.D. McKee. U.S. Geol. Survey Prof. Paper, No. 1052.

Wang, Y., Z. Lin, X. Lei, and S. Wang. (1987). Fabric and other physico-mechanical properties of loess in Shaanxi Province, China. In *Loess and Environment* (pp. 1–10). Ed. by M. Pecsi. Catena Supplement 9.

Wen, Q., G. Diao, and S. Yu. (1987). Geochemical environment of loess in China. In *Loess and Environment* (pp. 35–46). Ed. by M. Pecsi. Catena Supplement 9.

Wilding, L.P., Bui, E., Yerima, B., Pfordresher, B., Puentes, R., Ouattara, M., and Manu, A. (1989). Chemical and physical characteristics of dust inputs in the Sahel. In *TropSoils Technical Report, 1986–1987* (pp. 347–349). Ed. by N. Caudle and C. Levesque. Raleigh: North Carolina State University.

Windom, H.L. (1969). *Geol Soc. Am. Bull.*, 80:761–782.

4 Fluvial Systems

INTRODUCTION

Intensive cultivation of fluvial materials is common throughout the world. These areas also have complex vertical and horizontal changes in depositional environments and sediments. The soil pattern is simple in the middle of a large backswamp environment or can be very complex in areas close to the channels. Soil variability can range from minor to extreme depending upon the areal association of depositional environments. Constructing and describing suitable soil map units in fluvial materials is difficult.

Soil scientists probably work more with fluvial deposits or fluvially modified landscapes than any other material. Fluvial materials dominate many landscapes, and residual landscapes are no exception. Many upland regions have extensive areas of soils developed wholly in materials moved and sorted by the combined effects of water and gravity (see Chapter 5). Each upland soil map unit in the southern Piedmont has 25 to 35% of its area in head and foot slopes. These positions have a thin to moderate thickness of fluvial materials (Daniels et al., 1985). Investigations by geologists show that nearly 50% of the Southern Piedmont surface has a cover of mostly fluvial or water-worked material (Eargle, 1940; Whittecar, 1985).

All rolling landscapes have considerable area of valley bottoms and lower slopes covered by fluvial materials. Even undissected late Wisconsinan till landscapes have an extensive cover of later fluvially modified materials (Burras & Scholtes, 1987; Walker, 1966). A knowledge of the various subenvironments within the fluvial system helps one understand the lateral and vertical distribution of sediments.

Major fluvial deposits are those from braided or meandering streams or from alluvial fans. These deposits are important soil-forming materials because they occur as major or minor bodies in most landscapes. Sediment size, load, flow velocity and stream power determine channel type (Fig. 4.1). Thus, one can predict the kinds of sediment within an area from channel patterns alone. Although fluvial materials are exceedingly complex, there is order and predictive value within the various depositional elements.

BRAIDED STREAM DEPOSITS

Braided streams occur near modern glaciers or in deserts where channel flow is intermittent. The stream channel has moderate to high slope and an abun-

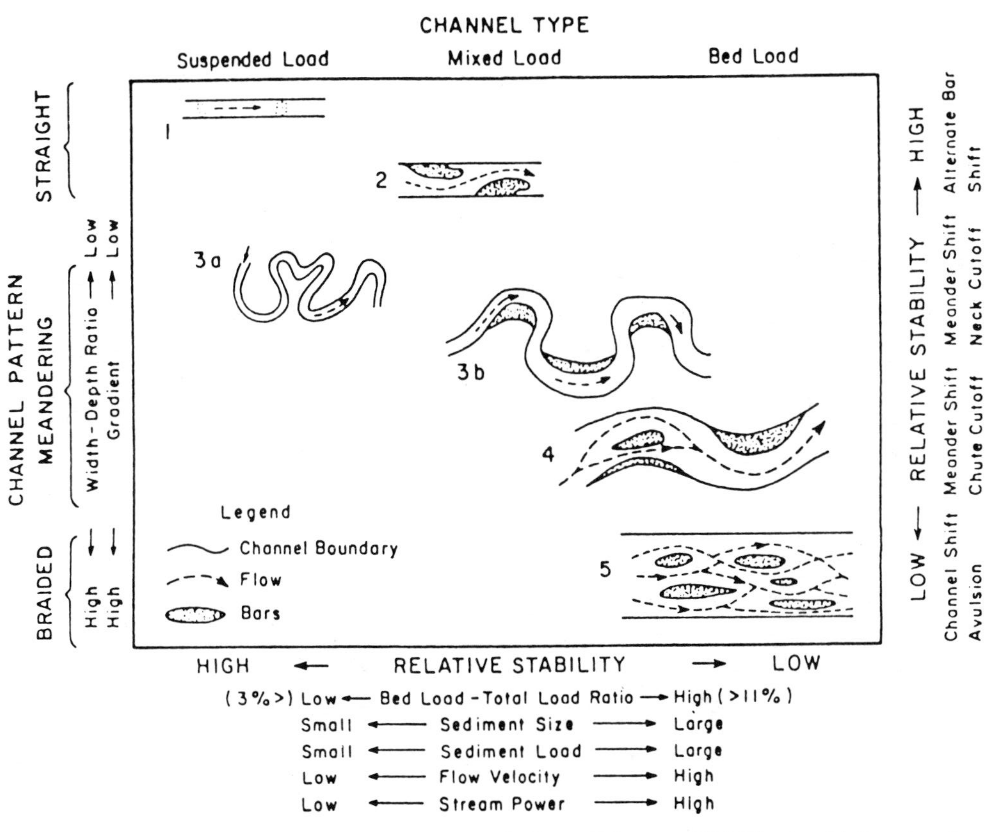

FIGURE 4.1. Classification of alluvial channels. From Schumm, 1981, Fig. 4, p. 24. Reprinted by permission from Society of Economic Paleontologists and Mineralogists.

dant source of sediment. The rapidly shifting channels deposit the sediment load of sand and gravel in bars. The bars and channels are in a nearly constant state of flux.

Particle size decreases downstream, although braided stream deposits are mostly sand and gravel that accumulate under shallow flows (Galloway & Hobday, 1983; Lewis, 1984; Miall, 1977). The energy of the system prevents deposition of silt and clay except in waning stages of flow. Sedimentation starts after scour concentrates a lag of coarse debris. Fine-grained materials settle out as current velocity decreases and produces a fine-grained top of the sequence. The next flood usually removes both the fine-grained and coarser material. The pools have deposits of mud and silt (Miall, 1977). The fine-grained top of braided streams is a major source of loess and eolian sands between floods or in the autumn when stream-flow volume decreases. The waning stages of flow deposit these materials and they are available for wind transport when they dessicate on an unvegetated surface (Smalley, 1975). Figure 4.2 is a braided stream model from Williams and Rust (1969). Table 4.1 shows the major coarse facies.

The areal extent of braided deposits may be large near or in glaciated regions and in some deserts. Few soils develop directly in coarse braided sediments. Most soils form in a finer texture material that overlies braided stream sediments with only the lower soil horizons in coarse material.

MEANDERING STREAM DEPOSITS

Environments

Alluvial plains are surfaces of low relief underlain by sediments deposited by rivers. These plains have several subdivisions (Lewis, 1984), including tributary fans, natural levees, point bars, splays, backswamps, abandoned channels and valley floor. Friedman and Sanders (1978) divide the sediments of flood plain rivers into channel, channel-marginal and overbank deposits. Channel deposits grow by lateral sedimentation and most overbank deposits grow by vertical accretion. Figure 4.3 is a meandering stream model (Galloway, 1977; Selley, 1982).

Channel deposits (Figs. 4.4 and 4.5) change properties rapidly, both vertically and horizontally. The coarse base includes lags of boulders, gravel, logs, bones and mud balls (Reineck & Singh, 1980). These coarse sediments grade upward into finer sands, accompanied by a change in the sedimentary structures (Fig. 4.4).

Point bar deposits have arcuate patterns in association with meander scars. They are the coarsest material normally exposed at the surface in fluvial sediments and are a major source of sand and gravel. A layer of finer overbank materials usually covers point bar deposits in high flood plains or terraces. The finer sediments probably are overbank deposits or natural levees formed as the channel shifts laterally.

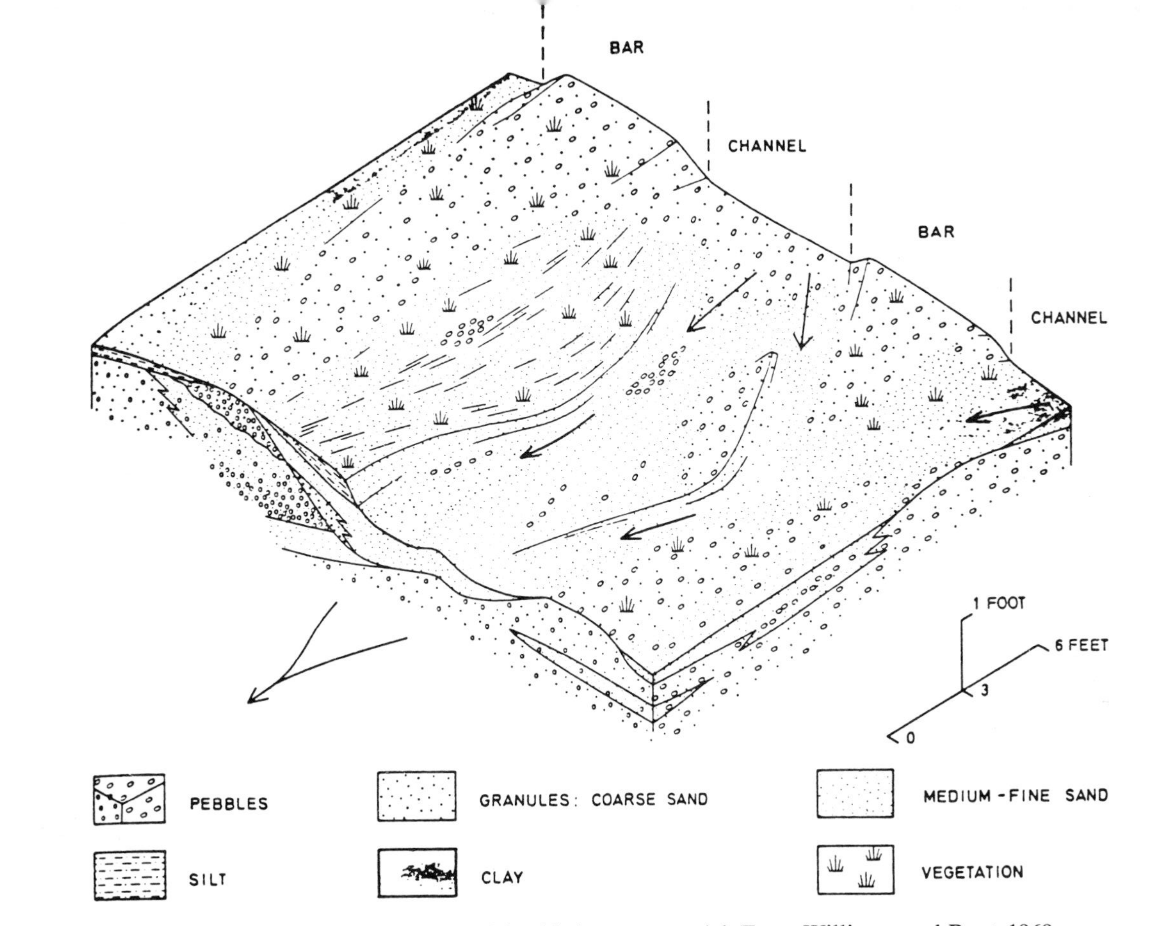

FIGURE 4.2. A three-dimensional braided stream model. From Williams and Rust, 1969, Fig. 19, p. 71. Reprinted by permission from the Society of Sedimentary Geology.

TABLE 4.1. Lithofacies for Six Principal Facies Assemblages of Braided Stream Deposits

Lithofacies	Structures	Interpretation
Massive matrix supported gravel	None	Debris flow deposits
Massive matrix supported or crudely bedded gravel	Horizontal bedding, imbrication	Longitudinal bars, lag deposits, sieve deposits
Stratified gravel	Trough cross-beds	Minor channel fills
Stratified gravel	Planar cross-beds	Linguoid bars or deltaic growths from older bars
Medium to coarse sands	Trough cross-beds	Dunes (lower flow regime)
Very fine to coarse sands	Ripple marks	Ripples (lower flow regime)
Very fine to coarse sands (may be pebbly)	Horizontal lamination	Planar bed flow
Fine sand	Low angle cross-beds (<10°)	Scour fills, crevasse splays

From Lewis, 1984, Table 2, p. 3.

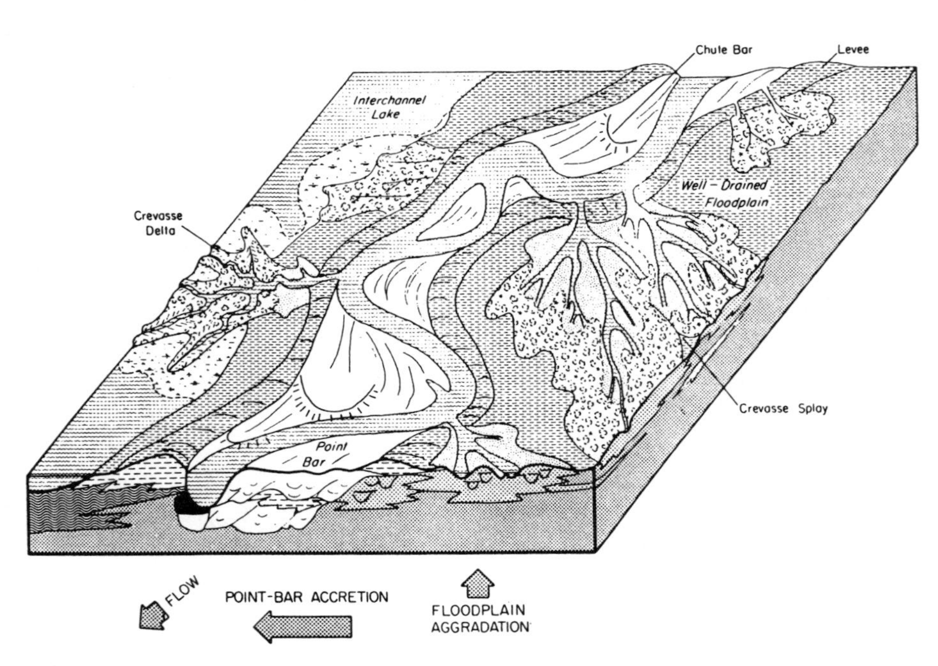

FIGURE 4.3. Meandering stream model. From Galloway, W.E., 1977, Fig. 13, p. 22. Reprinted by permission from W.E. Galloway.

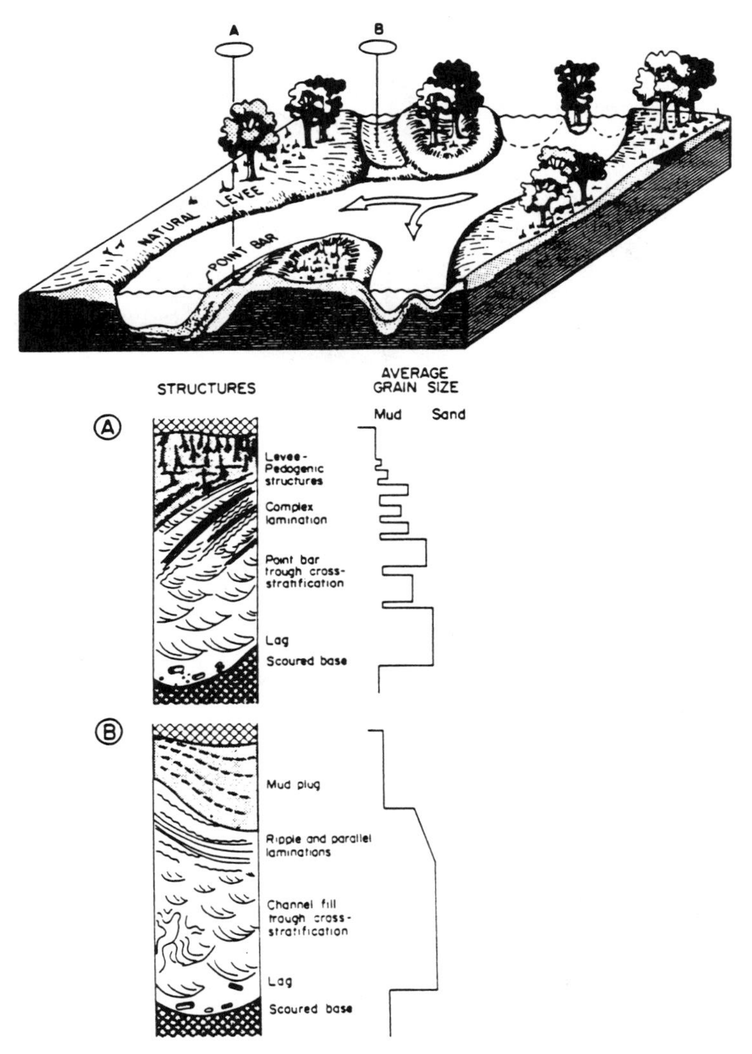

FIGURE 4.4. Generalized depositional model and vertical sequences of an anastomosed channel system. From Galloway and Hobday, 1983, Fig. 4-5, p. 58. Reprinted by permission from Springer-Verlag and W.E. Galloway.

Natural levees (Fig. 4.5) and crevasse splays (Fig. 4.6) are channel-marginal deposits (Fisk, 1947). Natural levees build upward where flood waters spill out in broad, shallow sheets. They occur only along channels carrying sand and silt in suspension and produce a laminated sheet deposit composed largely of fine sand to silt. Natural levees are common features in large river systems such as the Mississippi, but in the smaller streams they may be absent or subdued.

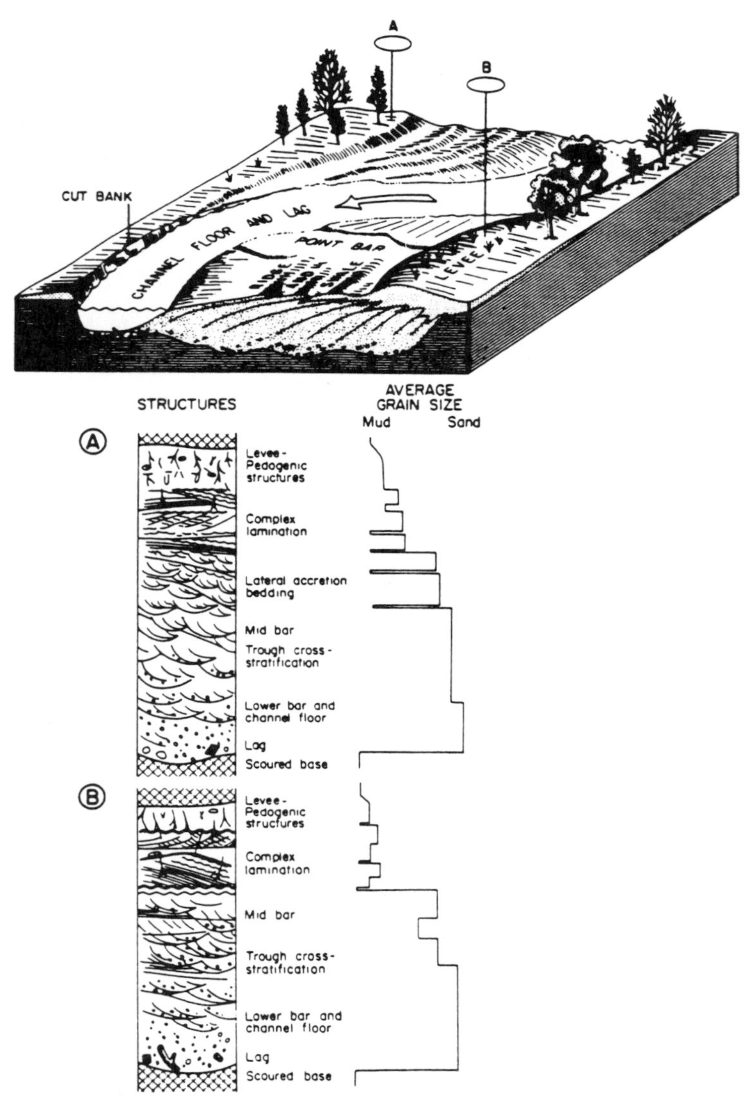

FIGURE 4.5. Generalized depositional model and vertical sequences of a high-sinuosity channel. From Galloway & Hobday, 1983, Fig. 4-6, p. 60. Reprinted by permission from Springer-Verlag and W.E. Galloway.

Floods erode gaps in levees and develop lobate crevasse splay deposits (Fig. 4.6). Splays have textures similar to or slightly coarser than the natural levee. Mean sediment diameter of splays decreases away from the eroded gap. Texture of the splays depends upon the energy of the environment and the size of material in the system at the time of flooding. Crevasse splays are prominent features on aerial photographs of cultivated fields and can occupy large areas in major river systems. Crevasse splays may be absent in small river systems. Flood channels or other sediments deposited during flood periods can interrupt the usual sequence of materials away from the active channel. Flood channels can have a wide range of textures depending upon source area composition and the energy of the flooded environment.

Overbank materials form beneath slowly moving or ponded water over sediments of other flood plain environments (Friedman & Sanders, 1978; Selley, 1982). The slowly moving water deposits a sheet or blanket of fine sand to silty clay sediment. Ponded areas of the flood plains or backswamps have fine-grained deposits that form pockets or lenses within the overbank silt. In large stream systems, the backswamps can cover broad areas and develop thick clay deposits.

Other sediments of alluvial plains include clay plugs of abandoned channels and colluvium from adjacent valley walls. Fans and deltas from tributaries increase the areal complexity of the sediments. Both tributaries and colluvium from the valley slopes interrupt the usual fining of deposits away from the controlling stream.

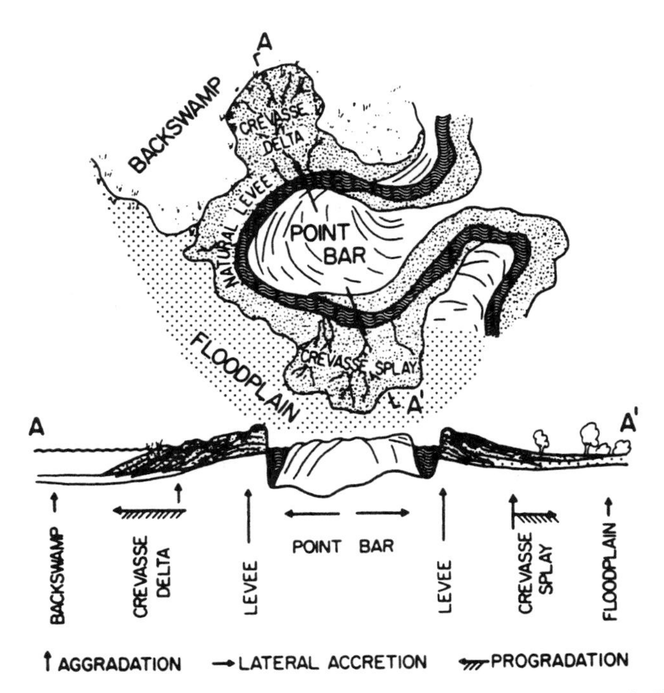

FIGURE 4.6. Relation of a crevasse splay to other land forms. From Galloway and Hobday, 1983, Fig. 4-9, p. 65. Reprinted by permission from Springer-Verlag and W.E. Galloway.

Application

Applying the meandering stream model to many small stream systems may require major revision. In small stream systems, soil materials usually are coarsest in the channels and fine away from the channel. Some landforms, such as point bar deposits, natural levees and crevasse splays or backswamps, may not occur. The relationship between sediment size and stream energy still holds.

The meandering stream model helps us predict soil textures within any given alluvial system. Figures 4.7A and B illustrate depositional environment and changes in soil texture away from a major channel in the Mississippi River Valley, Concordia and Catahoula Parishes, Louisiana. The natural levee is a silt loam near the channel and a silty clay loam near the transition to quiet water deposits. The quiet or slowly moving water deposits are clays thicker than 1 meter (Martin, 1988). The northeastern part of Figure 4.7 has a ridge and swale topography typical of point bar deposits. The ridges have soils with silty clay B horizons that overlie silt loam levee deposits (Boyd, 1986). The swales are clays, or quiet water deposits. Both the ridges and swales probably overlie a thick section of point bar deposits.

An experienced worker uses landforms on fluvial plains to predict the surface soil texture and internal drainage on fluvial plains. Making reasonable maps would be impossible without this predictive value of fluvial landforms. For excellent block dimensional diagrams of major soil associations and their

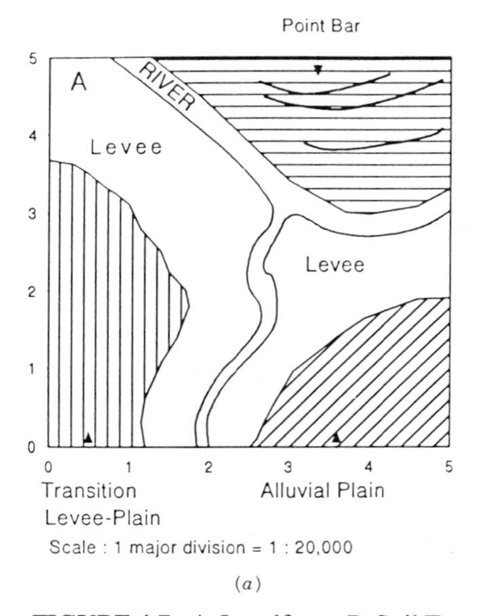

(a)

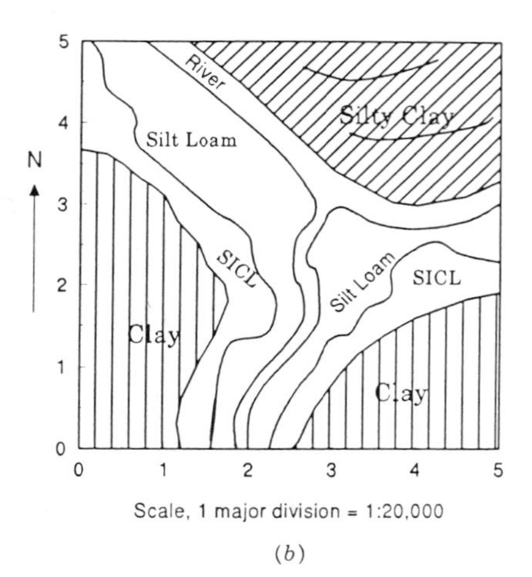

(b)

FIGURE 4.7. A. Landform. B. Soil Texture. From Martin, 1988, sheet no. 2, and Boyd, 1986, inset sheet 59.

relationship to sedimentary environment in the lower Mississippi River valley see Schumacher et al. (1988).

Map Unit Variability Flood plain soil map units have high textural and taxonomic variability. The study by Young et al. (1991) illustrates the textural heterogeneity within one relatively simple flood plain landform. The study used the Eudora soil map unit on old terraces of the Missouri River flood plain near Columbia, Missouri. The Eudora soil (coarse silty, mixed, mesic Fluventic Hapludolls) form in a silty veneer over sandy levee deposits. Most pedons have highly stratified C horizons and an A-C1-C2-C3 horizon sequence.

Transects of 12 Eudora map units selected at random, sampled at 10 locations at 200-foot intervals within each transect, gave a total of 120 sample sites. Laboratory analyses of the C1-C2 and C3 horizons included sand, silt and clay fractions. The sands coarser than very fine sand are the fs plus category. The silt textural class includes silts and very fine sands.

Figure 4.8 shows the textural variability of the C3 horizon and should not be unexpected in this depositional environment. The 120 pedons sampled contain 28 taxonomic classes, including 5 subgroups. Only 30 of the 120 observations are in the taxonomic family of the Eudora Soil (Fig. 4.9). Twenty-six pedons occur within 11 taxonomic families that have abrupt (contrasting) textural discontinuities within the C horizons. These data illustrate the difficulty soil taxonomy has in accommodating the natural variability of soils within one simple fluvial landform.

Reasonable soil maps are difficult to make in dissected fluvial landscapes because the predictive value is low. Outcrop of clay beds or gravel beds can occur within almost any elevation range. There is little texture-landform relationship on eroded valley slopes. The reduced predictive value is from the abrupt vertical and horizontal textural changes the fluvial systems.

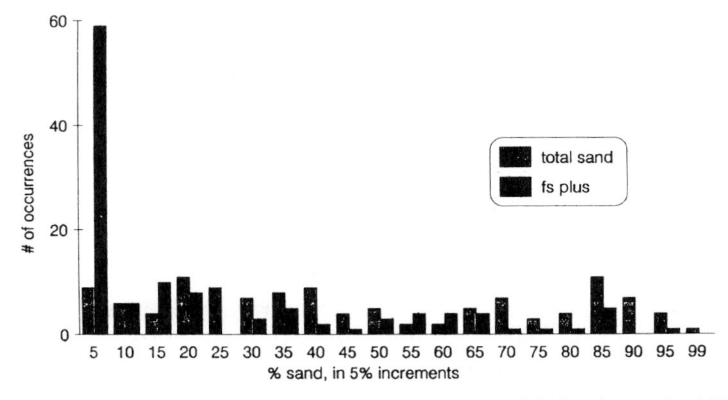

FIGURE 4.8. Frequency distribution of textural classes of C3 horizons for 120 pedons of the Eudora soil. From Young et al. (1991).

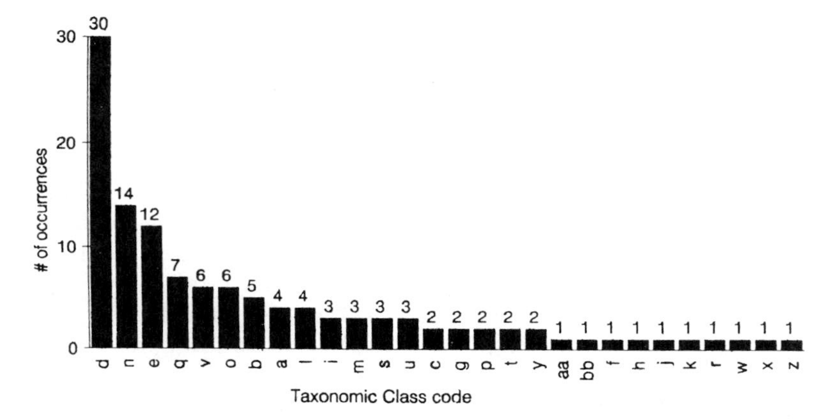

FIGURE 4.9. Frequency distribution of taxonomic classes for 120 pedons of the Eudora soil. A total of 28 families is represented.

Interpretations for Soils The typical fining upward of fluvial materials (Chorley et al., 1985; Leeder, 1982) also poses questions about interpretations of soil genesis. For example, thick soils in parts of the upper and middle Coastal Plain of North Carolina are in fluvial materials. The clay contents and mean diameters of these soils and sediments increase throughout the upper solum. The lower solum and C horizon have a nearly linear decrease with depth (Fig. 4.10; Table 4.2). Clay eluviation or destruction modified the upper part of the solum. Clay formation by weathering of primary minerals is possible in these soils, but only traces of feldspar and other weatherable minerals remain in the section (Gamble & Daniels, 1974). Thus, sedimentation processes probably are the primary control of the textures throughout the solum and C horizon. Material additions and losses during soil genesis in this type of deposit are difficult to measure. Soil map units in fluvial materials usually have considerable variability in lower solum properties.

FANS AND ASSOCIATED DEPOSITS

The arid and semiarid parts of the world have several unique depositional environments, according to some authors. Arid region landforms may differ somewhat from humid areas, but the depositional processes are similar. The environments in dry areas (Friedman & Sanders, 1978) are (1) bare rock surfaces, (2) pediments, (3) fans, (4) intermittent streams, (5) dunes, (6) sabkhas, and (7) playas. Environments 1,2,3,4, and 5 above also can occur in humid areas. Fans are not so topographically distinct in humid areas.

Fans

Fans can occur in humid and dry areas but are more common in faulted areas (Davis, 1983; Denny, 1967). All fans develop where a sediment-laden channel has an abrupt reduction in stream slope and competence. Prominent fans also form where intermittent streams leave the constriction of small valleys and debouch onto large valley floors. In both examples, an abrupt reduction in flow velocity causes the stream to drop part of its load. The most prominent fans are those in tectonically active arid regions.

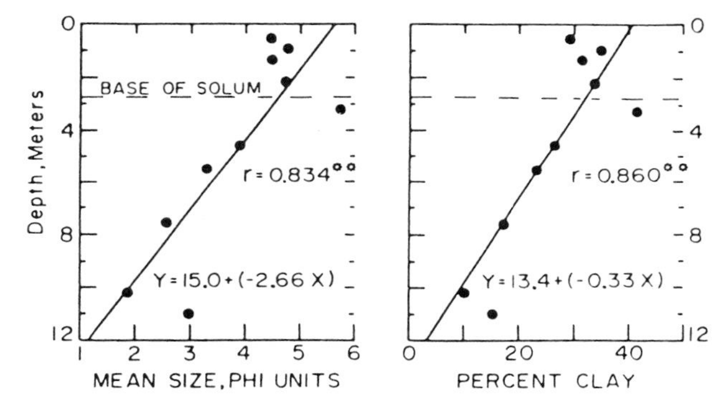

FIGURE 4.10. Clay distribution and mean size with depth of a soil formed in fluvial sediments. From Gamble and Daniels 1974, Fig. 2, p. 636.

TABLE 4.2. Particle Size Distribution of Soils and Underlying Sediments, Middle Coastal Plain, North Carolina

Depth (m)	Sand (%)	Silt (%)	Clay (%)	Mean Size (mm)
0.20– 0.30	77.9	15.7	6.4	0.13
0.46– 0.69	61.5	9.3	29.2	0.045
0.86– 1.07	58.0	7.3	34.7	0.036
1.27– 1.47	60.8	7.8	31.4	0.044
2.01– 2.39	60.2	5.9	33.9	0.037
		Base of solum		
3.05– 3.43	47.6	10.8	41.6	0.018
4.27– 4.88	66.1	7.2	26.7	0.067
5.18– 5.79	69.5	7.1	23.4	0.100
7.32– 8.84	74.7	8.0	17.3	0.17
10.06–10.36	82.2	7.5	10.3	0.27
10.36–11.58	73.2	11.5	15.3	0.13

Modified from Gamble and Daniels, 1974, Table 3, p. 636.

Environments A downslope change from bare-rock mountain slopes through fans to playas is common in arid region fans (Friedman & Sanders, 1978). The sequence of environments shown in Figure 4.11 exists in large areas of Arizona and New Mexico (Gile et al., 1981). Ideal arid fans have the shape of cone segments that radiate downslope, are concave upward, and have slopes of 1 to 25 degrees. Figure 4.12 is an idealized three-dimensional view of a fan.

Fans have a complex history of development with deposition, trenching and burial or abandonment as the area of deposition shifts (Figs. 4.12 and 4.13a,b,c; Denny, 1965; Harvey, 1989, Fig. 7.1). Particle size of the deposit decreases with decreasing slope and increasing distance from the apex (Table 4.3; Fig. 4.14). Still, any section may coarsen upward because the energy of the system varies considerably. Most fan properties such as slope, clast size and mean sediment diameter depend upon lithology, relief and climate of the contributing area (Gloppen & Steel, 1981; Ruhe, 1967).

Arid region fans are not simple systems. Channels shift easily, and new deposits and channels can differ considerably from the earlier ones (Figs. 4.12 and 4.13). Sediment particle size changes progressively from head to foot in arid fans (Fig. 4.14), although abrupt changes can occur. The three major mechanisms involved in sediment transport on arid region fans are stream flow, debris flow and mud flow (Hooke, 1967).

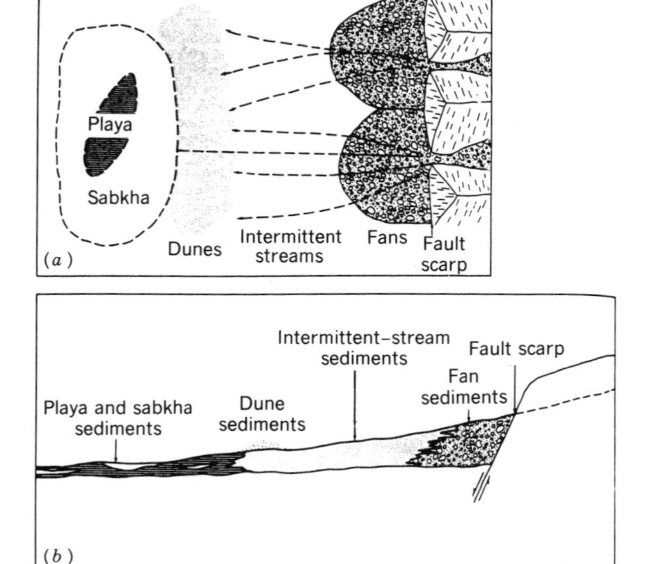

FIGURE 4.11. Areal relationship of bare rock surfaces, fan sediments and playa environments. From Friedman and Sanders 1978, Fig. 8-5, p. 203. Reprinted by permission from John Wiley Inc.

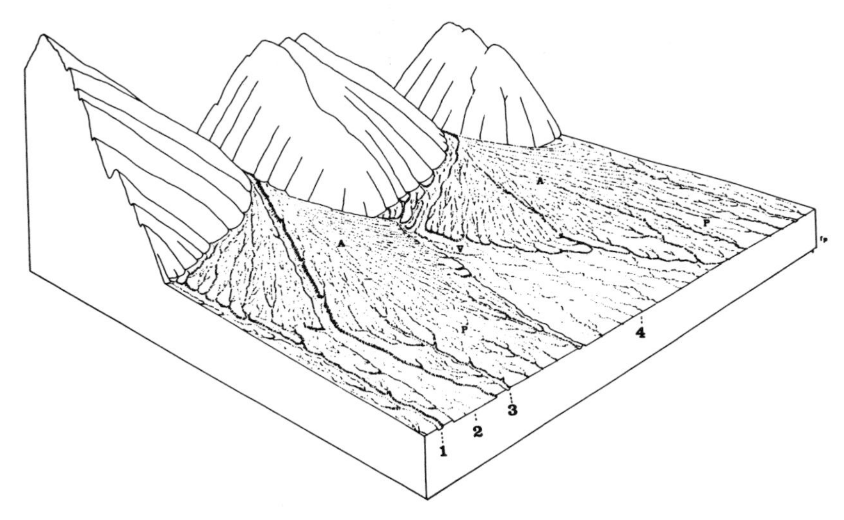

FIGURE 4.12. Idealized three-dimensional view of an arid region fan. From Peterson, 1981, Fig. 7, p. 61.

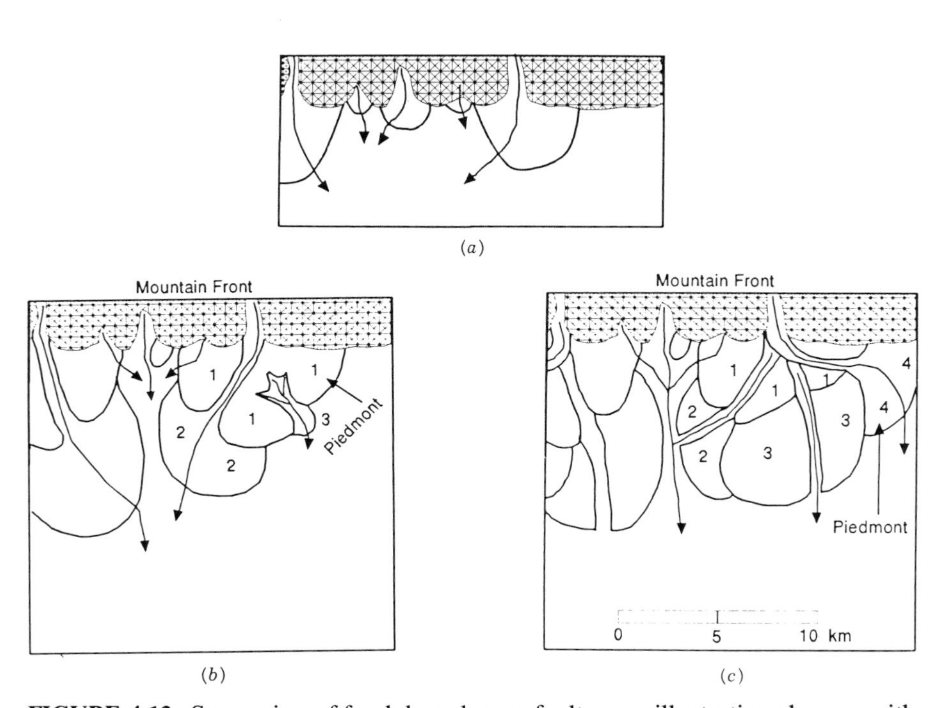

FIGURE 4.13. Succession of fan lobes along a fault scarp illustrating changes with increasing time; A is the initial deposition, C the latest deposition and reworking. From Denny, 1967. Reprinted by permission from *American Journal of Science* and C.S. Denny.

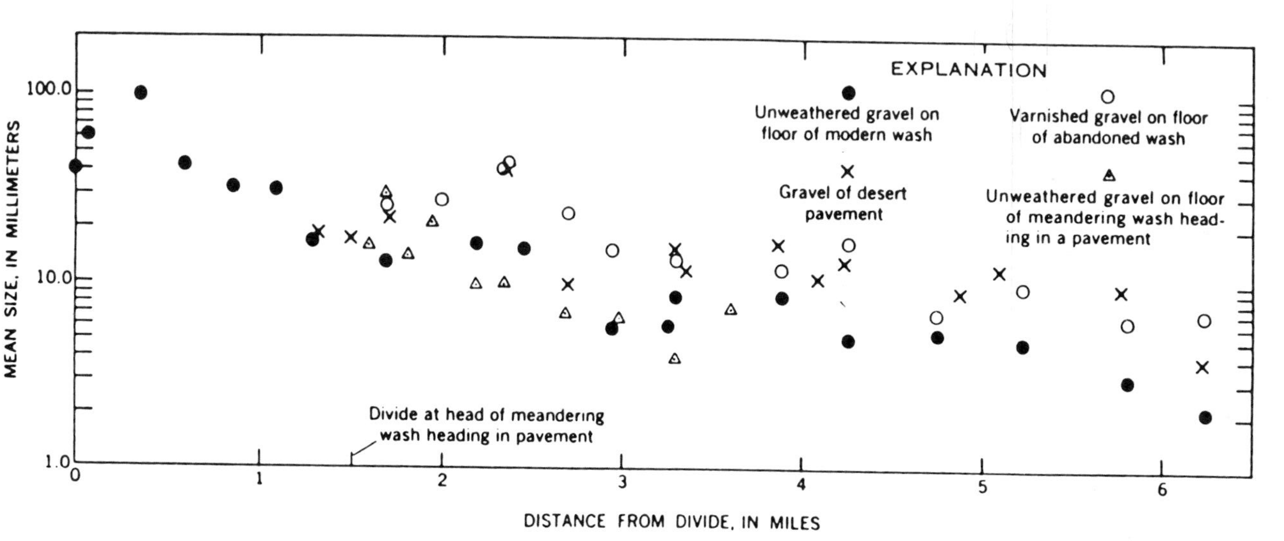

FIGURE 4.14. Changes in gravel size with distance from the divide. From Denny 1965, Fig. 5, p. 11. Reprinted by permission from C.S. Denny.

TABLE 4.3. Changes in Mean Particle Size Relative to Distance from a Mountain Front

Jornada		Organ	
Distance (miles)	Median Dia. (mm)	Distance (miles)	Median Dia. (mm)
0.50	1.35	0.25	1.70
2.00	0.85	0.60	2.00
3.75	0.05	1.50	1.35
3.90	0.11	2.65	1.90
4.65	0.11	3.60	1.25
5.00	0.17	4.90	0.82
5.75	0.12		
6.00	0.18		

From Ruhe, 1967, Table 7, p. 44. Reprinted by permission from the New Mexico Bureau of Mines and Mineral Resources.

Stream flow emerging from the retaining uplands spreads sheets of silt, sand and gravel over the fan. The flow deposits have little clay. Braided channels continually fill with well-sorted sediment and shift laterally (Galloway & Hobday, 1983). As with most braided stream deposits, sediment of individual flows fines upward in response to decreased velocity. The next flow erodes the fine upper materials unless the channel moves.

Debris flows are plastic masses that form when water moving across the fan mixes with sediment or loses water (Hooke, 1967). The poorly sorted or nonsorted debris flow sediments have cobbles and boulders set in a fine-grained mud matrix. Another stream flow may remove the mud matrix. Mud flows are similar to debris flows except they have sand-size or finer matrices. The mud flow is a muddy sand or a sandy mud and usually has 10% to 30% silt and clay. Mud flows are common in semiarid regions where glacial and volcanic materials are available.

Sieve deposits form by downstream loss of water (Galloway & Hobday, 1983; Hooke, 1967). They are lobes of coarse sediment stranded as water moves into a permeable gravelly substrate (Fig 4.15). Sieve deposits occur at mid-fan, and being channel system deposits, they are subject to reworking. Stranding followed by backfilling results in lenticular sieve deposits banked against an initial obstruction. A granitoid source that weathers to coarse gravel and sand combined with a high infiltration rate aids in the formation of sieve deposits. Many workers do not separate sieve deposits from debris flows or mud flows.

Figure 4.16 shows the areal distribution of the various deposits on a fan, but not all will be similar. The proportion of stream flow, debris flow and mud flow deposits in fans varies with the amount and frequency of rainfall and with watershed properties. Mud and debris flow deposits are abundant in arid region fans and stream flow deposits are common in humid region fans.

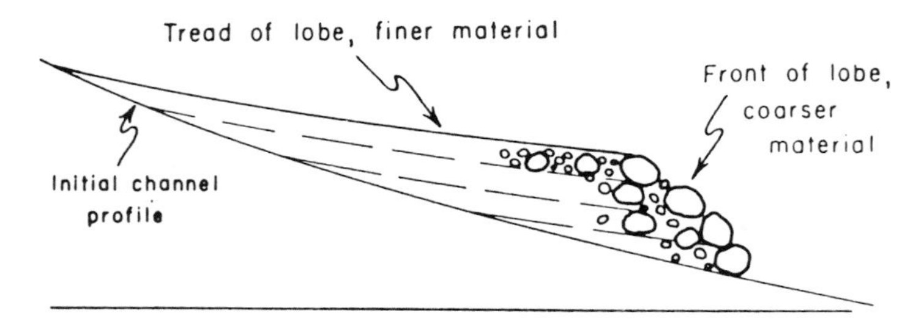

FIGURE 4.15. Idealized section of a sieve deposit showing the relationship of the coarser frontal lobe to the finer deposit beneath the tread. From Hooke, 1967, Fig. 9, p. 454. Reprinted by permission from University of Chicago Press.

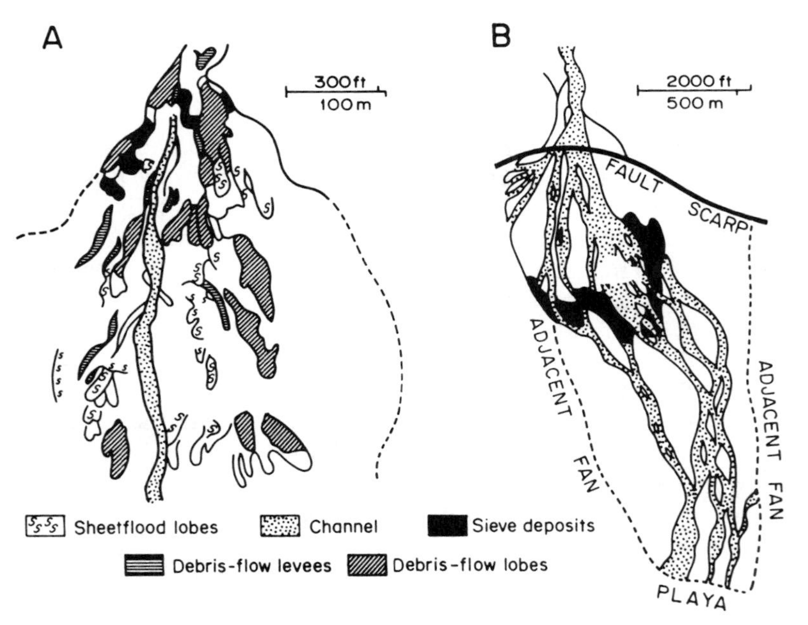

FIGURE 4.16. Areal distribution of fan sediments. From Galloway and Hobday, 1983, Fig. 3-4, p. 30, after Hooke, 1967. Reprinted by permission from Springer-Verlag, W.E. Galloway and University of Chicago Press.

Fluvial deposits dominate the sediments between the fans and the adjacent playa, and interfinger with the playas (Fig. 4.12). These deposits are similar to those of the fans, and many are debris flows. Fine-grained sandstones, siltstones and shales are uncommon in these deposits if their source is high-relief arid areas.

Playas The basin floor of arid basin and range areas of the U.S., commonly called a *bolson* (Peterson, 1981), can have both alluvial flats and playas. The alluvial flats are nearly level, graded surfaces underlain by sediments deposited by sheet floods or braided ephemeral streams. They are usually the most extensive areas in bolsons and have discontinuous eolian and lacustrine deposits.

Playa means a shore or bank of a body of water. Playa lakes are ephemeral because evaporation rates are high. The fine-grained sediments in playas are interbedded sand, silt, clay and evaporite minerals. Peterson (1981) defines basin and range playas as having four characteristics. They are

1. located on the floor of an intermountain basin,
2. have no vegetation,
3. flood ephemerally, and
4. have a surface veneer of fine-textured stratified sediments.

Playas may or may not be saline, but the surface usually crusts during dry periods.

Figure 4.17 shows the areal extent of major environments in or near a playa. Eolian sediments are common near playas and on fans. They may be continuous dune fields, coppice dunes or water reworked dune sands in the sabkhas (Friedman & Sanders, 1978; Gile et al., 1981). The changes in environment can be abrupt and areally complex.

Wet fans, fans in humid areas (Boothroyd, 1972), actively respond to monsoonal floods or ice melting within the watersheds. High annual precipitation, 1500 to 2500 mm, is typical, and many wet fans have perennial discharge (Galloway & Hobday, 1983; Schumm, 1977). Fans of humid areas usually have low gradients and are much larger than arid region fans, up to 16,000 km^2 (McGowen, 1979).

Fluvial processes dominate humid region fans, although debris flows can occur in the steeper areas of the upper fan. Clast size decreases downstream (Fig. 4.18; Boothroyd, 1972) with the lower fan becoming sandy (McGowen & Groat, 1971). Fine-grained deposits overlie coarser materials in the low-gradient parts of humid region fans. Sediment texture changes progressively within the fan system, as it is in arid region fans. Deposition rates on wet fans are as large as 15 meter in 2 to 3 years (Kuenzi et al., 1979).

Prediction Value

Particle size decreases with distance from the apex in fan and associated deposits. But in some arid region fans, no clear relationship exists between slope and clast size. Sediment composition depends upon the source area (Ruhe, 1967, 1975). Several contrasting environments can occur in short horizontal and vertical distances.

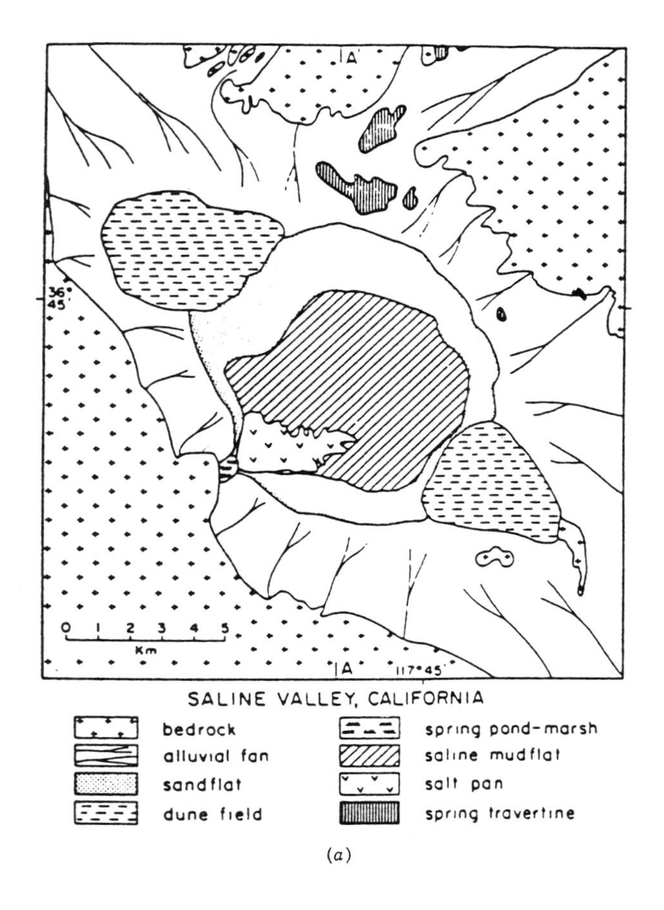

(*a*)

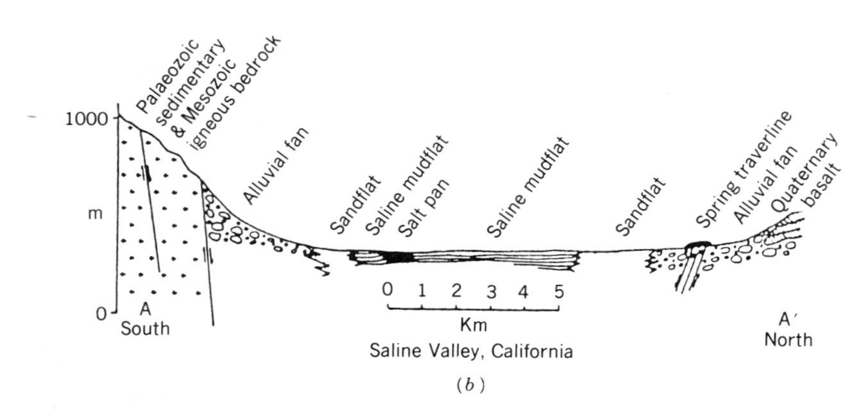

(*b*)

FIGURE 4.17. Areal extent of desert sedimentary environment. From Hardie et al., 1978, Fig. 5, p. 33. Reprinted by permission from the International Association of Sedimentationists.

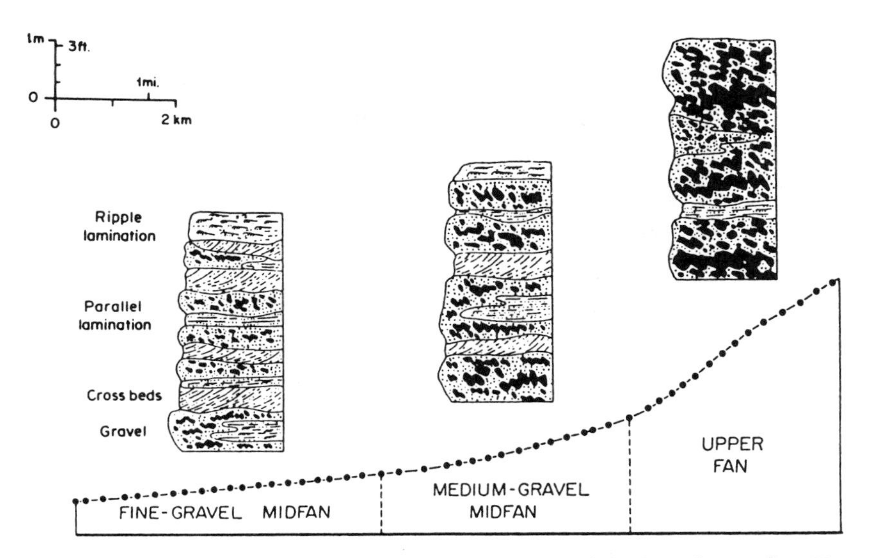

FIGURE 4.18. Downstream changes in gradient and particle size of a wet fan. From Boothroyd, 1972, Fig. 46, p. 122.

Thus, prediction of textural ranges at any point on a fan can be reasonably accurate. The specific texture at a point is only a guess because environments change abruptly. Later flows truncate fan sediments and expose a complex suite of environments and material textures. Soils formed on deeply dissected fans can have abrupt vertical and horizontal particle-size changes similar to those in dissected meandering stream deposits.

LAKES

Lakes are geologically ephemeral systems subject to filling from terrigenous sediments, evaporites or organic materials (Dean, 1981; Hakanson & Jansson, 1983). They are low-energy environments, and physical, chemical or biological processes dominate lake sedimentation (Rust, 1982). Local water and climate control the properties of lake sediments. Physical deposition of sediments dominates high-latitude and mountain lakes, and lakes with high-relief watersheds. High-latitude and mountain lakes have low chemical and biological activity.

Both the kinds and abundance of plants, continuity of cover in the watershed, and local relief affect the sediment yield. Chemical processes can dominate the ephemeral saline lakes, although sediment deposition also can have a major role. Biological activity dominates in some shallow eutrophic lakes (Rust, 1982). Water circulation, kinds and amounts of dissolved solids and temperature affect the first-cycle organic matter and chemical precipitation. Lake sediments range from gravel to organic materials to evaporite minerals. Positions within the lake, watershed composition and climate influence lacustrine deposits.

Shore and Shoreline Features

Shore and near shore features associated with lakes are eolian sand dunes, beaches and spits, and deltas and fans (Galloway & Hobday, 1983). Lakes with sandy beaches usually have some eolian sands. The lake size makes little difference because dunes occur next to large and small water bodies. Ice-rafted debris from thermal ice expansion or wind pushing ice to the lee shore is another lake margin deposit (Hobbs, 1911; Miller, 1971; Nichols, 1961; Pessl, 1969).

Freshwater Lakes Dominated by Sediment Input

Inflowing rivers control the physical sedimentation in lakes. Deposition of most of the fluvial load is near the entry point as deltas or fans. Lake deltas normally conform to the classic birdsfoot morphology because there is little dispersive activity by waves. Bedload deposits are near the entry point and currents carry suspended load for varying distances into the lake (Fig. 4.19).

Sediments are at the entry point and fine-grained in the lake center. Although sediments fine-outward into the basin, they coarsen-upward as the basin fills (Davis, 1983). The change in grain size away from the shoreline may change progressively from coarse to fine over a considerable distance (Fig. 4.17). Where the shore line is an erosional feature, the change from the coarse upland materials to fine lake sediments can be abrupt. The finest materials usually occur within the center of the lake where energy of deposition is lowest. Sediment mineralogy, organic content and sedimentation rate also change from the shore to the deeper areas (Picard & High, 1981).

Saline Lakes

Saline lakes, lakes with >5000 ppm dissolved solutes (Hardie et al., 1978), are common in arid regions. Tectonic setting and climates in which annual evaporation exceeds annual inflow control their distribution. Saline lakes may be perennial, such as the Great Salt Lake in Utah, or ephemeral such as those in many playas in the southwestern U.S.

Saline lake systems have a complex of related depositional subenvironments (Fig. 4.17; Shaw & Thomas, 1989, Figs. 9.2, 9.7) that control the texture and chemical properties of the sediments (Hardie et al., 1978). The chemistry and sediment textural relations are complex and cannot be discussed in this text in the detail needed. We suggest publications such as *Turbidites and Varves in Lake Brienz (Switzerland): Deposition of Clastic Detritus by Density Currents* (Strumm & Matter, 1978), and *Modern and Ancient Lake Sediments* (Matter &

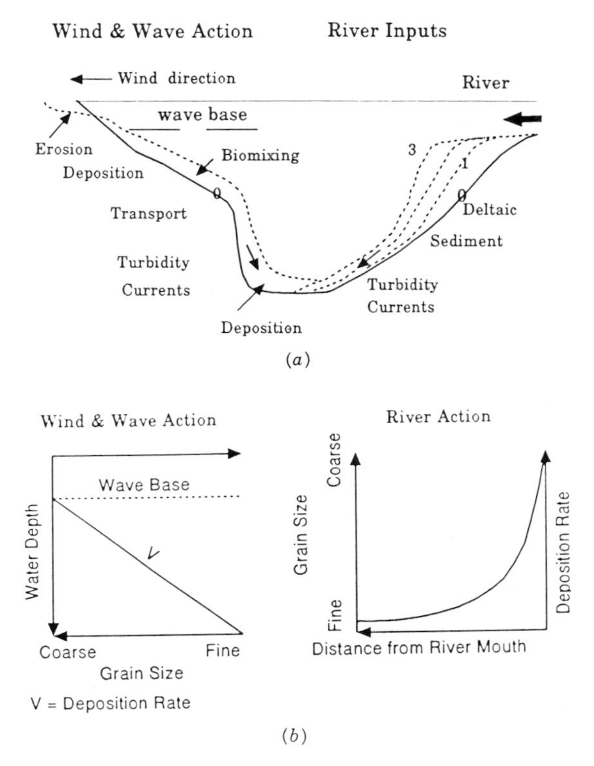

FIGURE 4.19. Bottom dynamics of lakes. From Hakanson, 1982, Fig. 1, p. 10. Reprinted by permission from Kluwer Academic Publishers.

Tucker, 1978). Eugster and Hardie's 1987 paper *Saline Lakes*, and *Evaporite Sedimentology* by Warren (1989) are both excellent references.

Recent publications that detail the relationships of salt-effected soils to location within the landscape are Skarie et al. (1986), Timpson and Richardson (1986), Skarie et al. (1987), Timpson et al. (1986), and Eghbal et al. (1989). The soils literature on this subject is extensive, but the above are good starting points.

Sediment Starved Lakes

The major deposit in sediment-starved lake basins is the organic materials that eventually fill the lake (Fig. 4.20). Organic materials in sediment-starved lakes are common from the Arctic to the equator. Such lakes range in size from small pot holes to large systems such as the Everglades.

Lacustrine organic sediments reflect the composition of the contributing area. The neutral organic sediments in the Florida Everglades occur only with

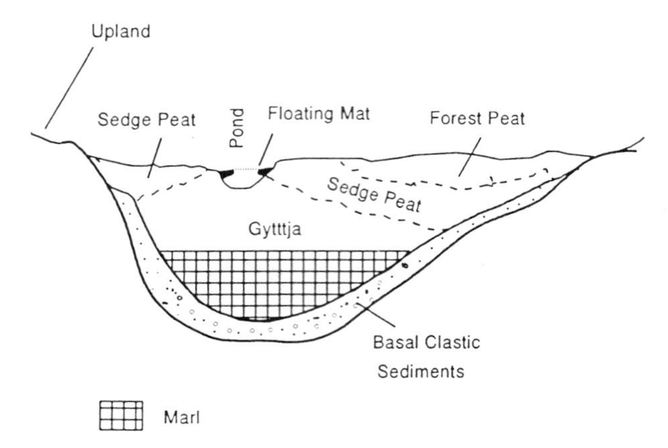

FIGURE 4.20. Organic materials in a sediment-starved lake. From Dean 1981, Fig. 2, p. 215. Reprinted by permission from Society of Economic Plaeontologists and Mineralogists.

limestone (McCollum et al., 1978). In Michigan, organic soils with neutral to alkaline reaction occur in areas draining calcareous till and outwash (Bowman, 1986). By contrast, the extremely acid organic soils in the North Carolina Pocosins (Godwin, 1989) overlie noncalcareous sediments. These soils receive nutrients largely from rainfall, not from a mineral watershed (Dolman & Buol, 1967).

Lacustrine Sediments and Soils

Exposed lake plains can have a complex association of sediments and soil. Reworking of the lake bed by wind and streams results in abrupt vertical changes in sediments. Wind can develop local eolian sand deposits, and streams can dissect or deposit fluvial materials over older lacustrine deposits. This reworking modifies the usual sequence of lake deposits outward from the shoreline, and can locally result in complex stratigraphic relationships.

In most lakes, specific environments have a predictable sequence and variability of sediments (Fig. 4.21). Figure 4.21 shows the topographic and some sedimentological relationships found in large lake systems in part of North Dakota. Figure 4.22 is a soil map of part of the area in Figure 4.19. It shows the textural relations from the lake plain across a beach ridge to another lake plain.

On a large lake plain such as that in Figure 4.21, one would expect the deeper beds of the lake plain to go beneath part of the beach ridge. Considerable interfingering of materials occurs at the edges. Similar relationships between sediments and ephemeral (pot hole) lakes exist. Near pot hole lakes the changes are even more abrupt and the areas involved commonly are too small to map (Fig. 3.7).

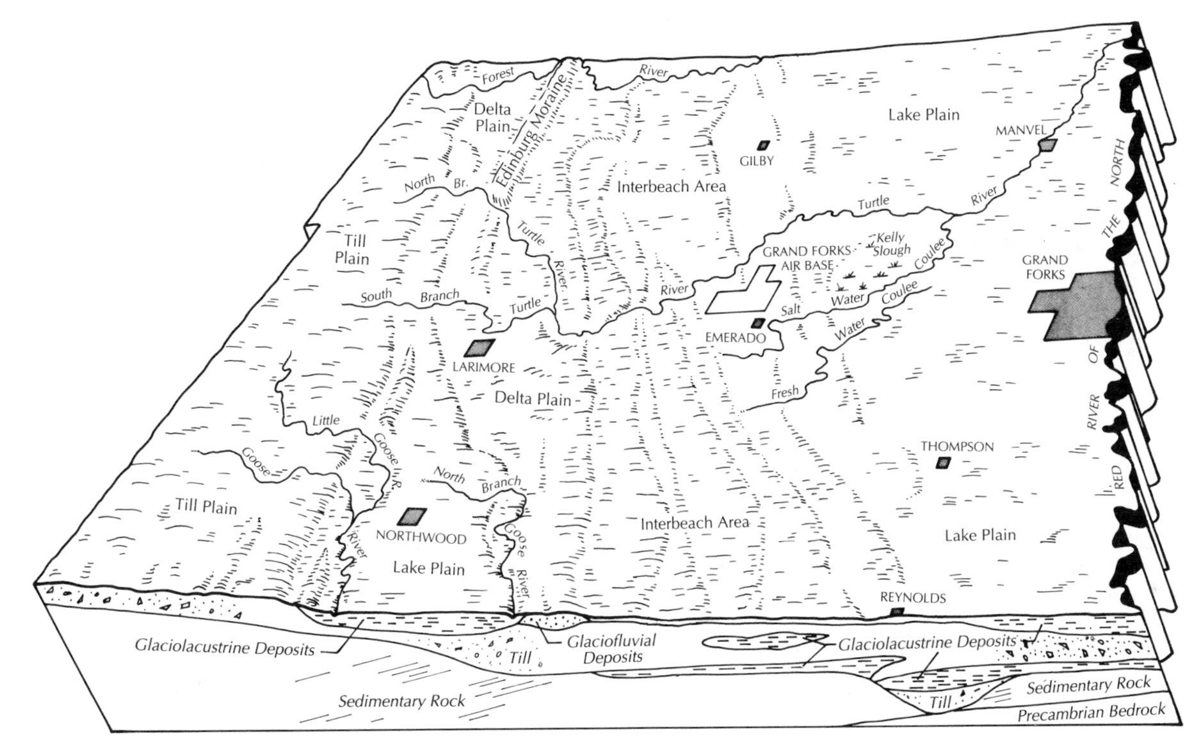

FIGURE 4.21. Idealized diagram of the physiography of Grand Forks County, North Dakota, showing the relationships between sediments and landscape. From Doolittle et al., 1981, Fig. 2, p. 3.

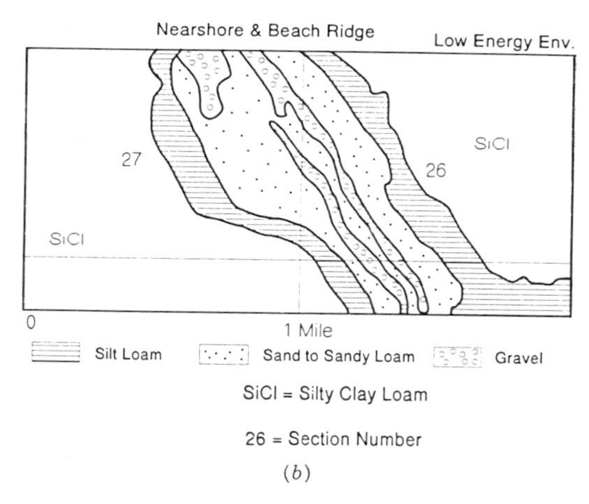

FIGURE 4.22. Sediment and soil changes from lake plain to beach ridge. A. Soil map. B. Soil and sediment textures. From Doolittle et al. 1981, soil survey sheet 107. *Soil map unit legend*: 4 Arveson loam; 35 Rauville silt loam; 48 Wyndmere sandy loam; 50B Hecla fine sandy loam; 54B Embden fine sandy loam; 67 Gilby loam; 97D Sioux loam; 126 Bearden silty clay loam; 171 Antler-Tonka silt loams.

ADDITIONAL READING

Allen, P.A. and J.D. Collison. (1986). Lakes. In *Sedimentary Environments and Facies* (pp. 63–94). Ed. by H.G. Reading. Boston: Blackwell.

Collison, J.D. (1986). Alluvial sediments. In *Sedimentary Environments and Facies* (pp. 20–62). Ed. by H.G. Reading. Boston: Blackwell.

Fisher, W.L. and L.F. Brown, Jr. (1984). *Clastic Depositional Systems—A Genetic Approach to Facies Analyses*. Austin: Bureau of Econ. Geol., University of Texas at Austin.

Fraser, G.S. (1989). *Clastic Depositional Sequences*. Englewood Cliffs, NJ: Prentice Hall.

Miall, A.D. (1985). Architectural element analysis: A new method of facies analysis applied to fluvial deposits. *Earth Science Reviews*, 22:261–308.

McPherson, J.G., Shanmugan, G., and Moiola, R.J. (1987). Fan-deltas and braid deltas: Varieties of coarse-grained deltas. *Geol. Soc. Amer. Bull.*, 99:331–340.

Schreiber, B.C. (1986). Arid Shorelines and Evaporites. In *Sedimentary Environments and Facies* (pp. 189–228). Ed. by H.G. Reading. Boston: Blackwell.

REFERENCES

Boothroyd, J.C. (1972). *Coarse-Grained Sedimentation on a Braided Outwash Fan, Northeast Gulf of Alaska*. Univ. South Carolina, Coastal Res. Div. Tech. Rept. No. 6–CRD.

Bowman, W.L. (1986). *Soil Survey of Van Buren County, Michigan*. USDA Soil Conservation Service.

Boyd, W.H. (1986). *Soil Survey of Catahoula Parish, Louisiana*. USDA Soil Conservation Service.

Burras, C.L. and W.H. Scholtes. (1987). *Soil Sci. Soc. Amer. J.*, 51:1541–1547.

Chorley, R.J., S.A. Schumm, and D.E. Sugden. (1985). *Geomorphology*. New York: Methuen.

Daniels, R.B., J.W. Gilliam, D.K. Cassel, and L.A. Nelson. (1985). *Soil Sci. Soc. Amer. J.*, 49:991–995.

Davis, R.A., Jr. (1983). *Depositional systems*. Englewood Cliffs, NJ: Prentice-Hall.

Dean, W.E. (1981). *Carbonate Minerals and Organic Matter in Sediments of Modern North Temperate Hard-Water Lakes*. Int. Soc. Econ. Paleont. Mineral. Spec. Pub. 31:213–231.

Denny, C.S. (1965). *Alluvial fans in the Death Valley region, California and Nevada*. U.S. Geol. Survey Prof. Paper, No. 466.

Denny, C.S. (1967). *Am. J. Sci.*, 265:81–105.

Dolman, J.D. and S.W. Buol. (1967), *A Study of Organic Soils (Histosols) in the Tidewater Region of North Carolina*. North Carolina Agric. Exp. Sta. Tech. Bull., No. 181.

Doolittle, J.A. et al. (1981). *Soil Survey of Grand Forks County, North Dakota*. USDA Soil Conservation Service.

Eargle, D.H. (1940). *Science*, 91:337–388.

Eghbal, M.K., R.J. Southard, and L.D. Whittig. (1989). *Soil Sci. Soc. Amer. J.*, 53:898–903.

Eugster, H.P. and L.A. Hardie. (1987). Saline lakes. In *Lakes—Chemistry, Geology and Physics* (pp. 237–293). Ed. by A. Lerman. New York: Springer-Verlag.

Fisk, H.N. (1947). *Fine Grained Alluvial Deposits and Their Effects on Mississippi River Activity*. Vicksburg, MI: Mississippi River Commission.

Friedman, G.M. and J. E. Sanders. (1978). *Principles of Sedimentology*. New York: Wiley.

Galloway, W.E. and D.K. Hobday. (1983). *Terrigenous Clastic Depositional Systems*. New York: Springer-Verlag.

Gamble, E.E. and R.B. Daniels. (1974). *Soil Sci. Soc. Am. Proc.*, 38:633–637.

Gile, L.H., J.W. Hawley, and R.B. Grossman. (1981). *Soils and Geomorphology in the Basin and Range area of Southern New Mexico—Guidebook to the Desert Project*. New Mexico Bureau of Mines and Mineral Resources, Memoir 39.

Gloppen, T.G. and R.J. Steel. (1981). The deposits, internal structure, and geometry in six alluvial fan-fan delta bodies (Devonian-Norway)—a study in the significance of bedding sequence in conglomerates. In *Soc. Econ. Paleont. Mineral. Spec.*, [Pub. 31] (pp. 49–69).

Godwin, R.A. (1989). *Soil Survey of Craven County, North Carolina*. USDA Soil Conservation Service.

Hakanson, L. (1982). *Hydrobiologia*. 91:9–22.

Hakanson, L. and M. Jansson. (1983). *Principles of Lake Sedimentology*. New York: Springer-Verlag.

Hardie, L.A., J.P. Smoot, and H.P. Eugster. (1978). Saline lakes and their deposits: A sedimentological approach. In *Modern and Ancient lake Sediments* (pp. 7–42). Ed. by A. Matter and M.E. Tucker. [Int. Assoc Sedimentologist Sp. Pub. No. 2] London: Blackwell.

Harvey A.M. (1989), The occurrence and role of arid zone alluvial fans. In *Arid Zone Geomorphology* (pp. 136–158). Ed. by D.S.G. Thomas. New York: Halsted Press.

Hobbs, W.H. (1911). *J. Geol.*, 19:157–160.

Hooke, R.L. (1967). *J. Geol.*, 75:438–460.

Kuenzi, D.W., O.H. Horst, and R.V. McGehee. (1979). *Geol. Soc. Am. Bull.*, 90:827–838.

Leeder, M.R. (1982). *Sedimentology, Process and Product*. London: Allen & Unwin.

Lewis, E.W. (1984). *Practical Sedimentology*. Stroudsburg, PA: Hutchinson & Ross.

Martin, P.G. (1988). *Soil Survey of Concordia Parish, Louisiana*. USDA Soil Conservation Service.

Matter, A. and M.E. Tucker (Eds.) (1978). *Modern and Ancient Lake Sediments*. International Association of Sedimentologists, Special Publication No. 2.

McCollum, S. et al. (1978). *Soil Survey of the Florida Palm Beach Area*. USDA Soil Conservation Service.

McGowen, J.H. (1979). Coastal plain systems. In *Depositional and Ground-Water Flow Systems in Exploration for Uranium, a Research Colloquium* (pp. 80–117). Austin: Bur. Econ. Geol., Univ. Texas.

McGowen, J.H. and C.G. Groat. (1971). *Van Horn Sandstone, West Texas: An Alluvial Fan*

Model for Mineral Exploration [Rept. Invest. No. 72] Austin: Bur. Econ. Geol., Univ. Texas.

Miall, A.D. (1977). *Earth Sci. Review*, 13:1–62.

Miller, G.A. (1971). *A Geomorphic, Hydrologic and Pedologic Study of the Iowa Great Lakes Area*. Unpublished masters thesis. Ames: Iowa State University.

Nichols, R.L. (1961). *Amer. J. Sci.*, 259:694–708.

Peterson, F.F. (1981). *Tech. Bull. 28*. Nevada Agric. Exp. Sta..

Pessl, F. Jr. (1969). *Formation of a Modern Ice-Push Ridge by Thermal Expansion of Lake Ice in Southeastern Connecticut*. CRREL Res. Rept. 295.

Picard, M.D. and L.R. High, Jr. (1981). *Physical Stratigraphy of Ancient Lacustrine Deposits*. Soc. Econ. Paleont. Mineral. Spec. Pub. 31:233–259.

Reineck, H.E. and I.B. Singh. (1980). *Depositional Sedimentary Environments*, 2nd ed. New York: Springer-Verlag.

Ruhe R.V. (1967). *Geomorphic Surfaces and Surficial Deposits in Southern New Mexico*. New Mexico Bur. Mines & Min. Res. Memoir 18.

Ruhe, R.V. (1975). *Geomorphology*. New York: Houghton Mifflin.

Rust, B.R. (1982). *Hydrobiologia*, 91:59–70.

Schumacher, B.A., W.J. Day, M.C. Amacher, and B.J. Miller. (1988). *Soils of the Mississippi River Alluvial Plain in Louisiana*. Louisiana Agric. Exp. Sta. Bulletin No. 796.

Schumm, S.A. (1977). *The Fluvial System*. New York: Wiley.

Schumm, S.A. (1981). *Econ. Paleontologists and Mineralogists Spec. Pub*. 31:19–29.

Shaw, P.A. and D.S.G. Thomas. (1989). Playas, pans and salt lakes. In *Arid Zone Geomorphology* (pp. 184–205). Ed. by D.S.G. Thomas. Halsted Press.

Selley, R.C. (1982). *An Introduction to Sedimentology*. London: Academic Press.

Skarie, R.L. et al. (1986). *J. of Environ. Quality*, 15:335–340.

Skarie, R.L. et al. (1987). *Soil Sci. Soc. Amer. J.*, 51:1372–1377.

Smalley, I.J. (1975). *Loess Lithology and Genesis*. [Benchmark Papers in Geology, no. 20] New York: Wiley.

Strumm, M. and A. Matter. (1978). *Turbidites and varves in lake brienz (Switzerland): Deposition of clastic detritus by density currents*. In *Int. Assoc. Sediment.*, Spec. Pub. 2 (pp. 147–168).

Timpson, M.E. and J.L. Richardson. (1986). *Geoderma*, 37:295–305.

Timpson, M.E. et al. (1986). *Soil Sci. Soc. Amer. J.*, 50:490–493.

Walker, P.H. (1966). Postglacial environment in relation to landscape and soils on the Cary Drift, Iowa. *Iowa Agri. Exp. Sta. Res. Bull.*, 549:838–875.

Warren, J.K. (1989). *Evaporite Sedimentology*. Englewood Cliffs, NJ: Prentice Hall.

Whittecar, G.R. (1985). *SE Geol.*, 26:117–129.

Williams, P.F. and B.R. Rust. (1969). *J. of Sed. Petr.*, 39:649–679.

Young, F.J., R.D. Hammer, and J.M. Maatta. (1991). *Agron. Abstracts*.

5 Hillslope Sediments

INTRODUCTION

Hillslope sediments are thin deposits on valley slopes emplaced by the combined effects of gravity and water run-off from higher parts of the landscape. Many soil mappers call these reworked materials *colluvium*. Hillslope sediments can include the colluvial deposits mapped by geologists, but hillslope sediments usually are thinner.

These sediments (Fig. 5.1) occur on slopes ranging from 1 to 20%+ and are ubiquitous materials that occur in all erosional landscapes. Some geologists map hillslope sediments as a discontinuous cover of colluvium but many geologists and soil scientists may ignore or misinterpret hillslope sediments.

Soil scientists should recognize hillslope sediments because these materials have considerable influence on soil variability and moisture regimes (Butler et al., 1986). They also record geomorphic events that are important in determining site history (Butler, 1958, 1959; Kleiss, 1970; Ruhe et al., 1967). Usually, our first assumption upon examining a soil is that it formed in one material of a given age, not in multiple materials of variable ages. Recognizing multiple soil materials in a landscape makes the interpretation of field and laboratory data more complicated than a single material.

The recognition of hillslope sediments by careful observers is the beginning of field pedology. Milne (1936) and Robinson (1936) believed that surface creep and slow erosion were responsible for the textural zonation in certain areas. They concluded that some soil horizons formed from materials moved downslope by various processes. Eargle (1940) and Whittecar (1985) showed that about 50% of the southern Piedmont of the United States has a cover of erosional debris. Yet most of our soil surveys describe the upland map units as residual soils. Most surveys do not describe hillslope sediment derived from colluvial and slope wash processes.

In the 1950s and 1960s Australian and American soil scientists published several detailed reports about soils developed in more than one material (Bettenay & Hingston, 1964; Butler, 1958, 1959; Churchward, 1961, 1963; Jessup, 1961; Riecken & Poetsch, 1960; Ruhe, 1959; Ruhe et al., 1967; Van Dijk, 1959; Vreeken, 1968; Walker, 1962a, 1962b, 1966). Later, Kleiss (1970), Malo et al. (1974), Matzdorf et al. (1975) and McCracken et al. (1987) added additional detail to the properties and distribution of thin surficial materials and their influence on soil properties.

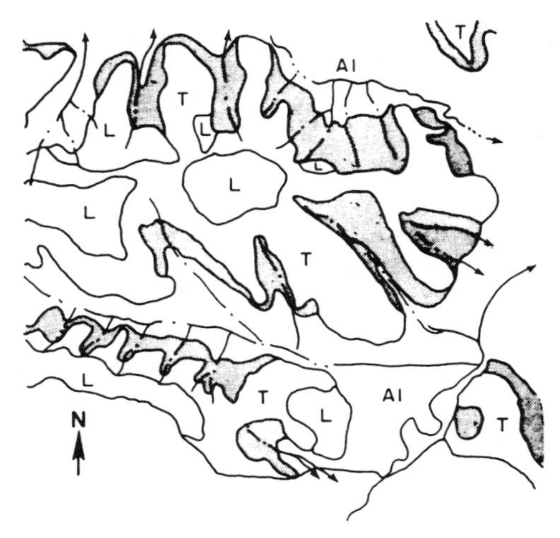

FIGURE 5.1. Areal extent of hillslope sediments (shaded areas) overlying glacial till. Redrawn from Ruhe et al., 1967, Plate 5.

Development of Soil Taxonomy required considerable effort in the late 1950s and 1960s (Soil Survey Staff, 1975). The new taxonomy marked a shift in philosophy from the previous taxonomic system that had placed emphasis on geologic deposits as a soil series criterion. Soil Taxonomy is a remarkable achievement and advanced our knowledge of soils systems. But nothing conceived by man is perfect: The taxonomic emphasis on quantitative soil classification resulted in a subtle but pervasive shift from interpretation of soils as components of landscapes to soils as units unto themselves.

Soil Taxonomy defines soils as single points in the soil continuum. There is little emphasis on the spectrum of integral relationships other than to the few select properties used in the classification. This shift is probably the result of deemphasis of materials and deposits as criterion for separating soils. The field soil scientist should apply as much effort to understanding the relationship of the pedon to the landscape as to classifying the soil. Accurate extrapolation of research data to the soil in its natural environment is as important as accurate laboratory procedures.

One example of the effects of our changing view of soils is the definition of a lithologic discontinuity. The original definition of a lithologic discontinuity was that it marked the boundary between two periods of material deposition (even if the materials are very similar). Two examples of a discontinuity are the contact between two loess sheets, or between till and an overlying loess. In both cases, there is a time interval between deposition, although the interval may be very small between till and loess. By decree, the pedologic definition of a discontinuity includes an abrupt change in texture. There is no requirement for a time interval between the deposition of the two contrasting materials.

A problem with the pedologic interpretation of a discontinuity is that a vertical change in texture has several interpretations for soil development. A silty or finer bed over sands and gravels is common in fluvial environments. Deposition of both beds occurred during a short time period, and they represent only a change in depositional environment. In a dissected loess-mantled landscape, the same sequence of sediments can occur on valley slopes and a different interpretation may apply. The interpretation is different if the finer bed is hillslope sediment derived from a loess mantle and overlies a truncated sand or gravel bed. Not only is the time interval greater than in the first example, but the geomorphic interpretation is much different.

The authors have participated in field trips during which textural breaks in stratified deposits (a clay loam bed over a sandy loam, for example) are lithologic discontinuities. Conversely, a mantle of fine sandy loam over a clay loam on a hillslope is not a discontinuity. This mantle is a pedogenic horizon developed by modification of the underlying material.

We question that on valley slopes the above interpretation is always correct. In most rolling landscapes, field evidence suggests the upper material has moved downslope. The relationships of the upper material to the surrounding landscape are important, but often ignored. With emphasis upon classification instead of evolution, we develop tunnel vision. We look only at holes in the ground instead of integrating the soil into its landscape.

DISTRIBUTION AND PROPERTIES

Hillslope sediments (defined by Kleiss, 1970) are those materials derived by erosional processes from upslope. The material may be from a combination of processes such as slope wash, faunal activities, creep or frost heave. In materials with gravel a lag concentrate, or stone line (Ruhe, 1959), usually occurs near or at the base of the recent deposit. The hillslope sediments are the same age as the alluvium they merge into at the foot or toe slopes (Walker, 1962a).

Distribution

Hillslope sediments are present in landscapes ranging from gently undulating late Wisconsinan till plains to dissected erosional topography (Burras & Scholtes, 1987; Hack & Goodlett, 1960; Kleiss, 1970; Walker, 1962a, 1966; Whittecar, 1985). They occur in various climatic and topographic regions and grade into alluvium on one end and mass movement deposits such as slumps on the other.

Only detailed work recognizes hillslope sediments as a stratigraphic unit. Yet they occur on almost all erosional landscapes. Recognition of these sediments is easy where gravel or buried soils mark the boundary, but they also occur in loess and other materials without gravel. In loess landscapes, the contact between loess and silty hillslope sediment is not visible; proof of its presence requires detailed laboratory data (Huddleston & Riecken, 1973).

Hillslope materials occur in most head slopes and some linear slopes in rolling erosional landscapes (Fig. 5.1). These materials are absent or thin on nose slopes and the upper part of the linear slopes (Fig. 5.2). Hillslope sediment forms a nearly continuous cover on virgin rolling landscapes (McCracken et al., 1987). On many rounded interfluves and linear slopes these sediments are thin and incorporated into the Ap horizon, which makes identification difficult. The sediments usually deepen downslope as they merge with the adjacent alluvial fill (Kleiss, 1970; Riecken & Poetsch, 1960; Walker, 1962a, 1962b).

Properties

Hillslope sediments thicken downslope and may cross several soils, or may be confined to one soil in the landscape (Fig. 5.3 a and b). The particle size of the material depends upon the material within its local watershed. For example, Kleiss (1970) found a constant decrease in hillslope sediment geometric mean diameter from the interfluve to midslope. A sand lens at midslope produced an

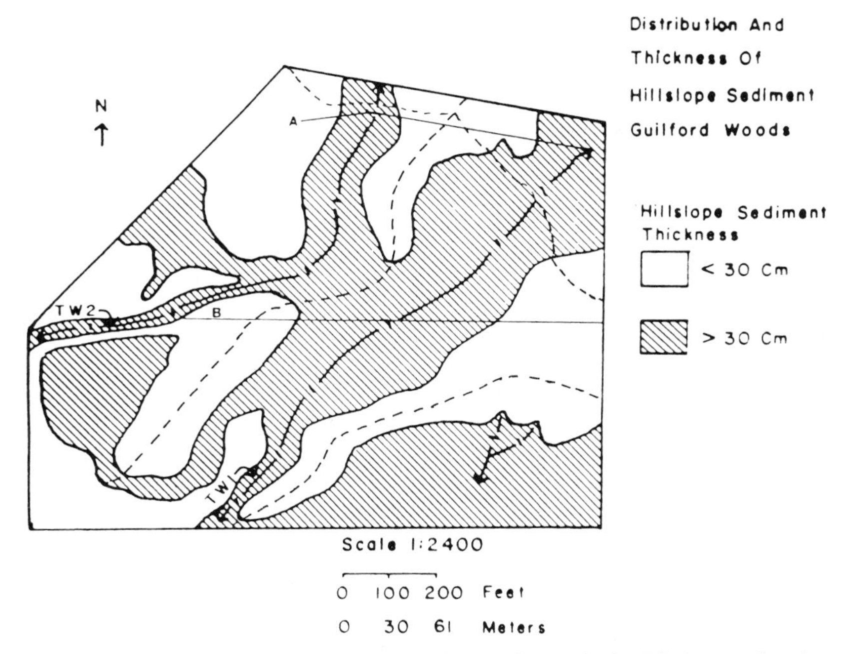

FIGURE 5.2. Distribution of hillslope sediment in a virgin Piedmont landscape. From McCracken et al., 1987, Fig. 1B, p. 92.

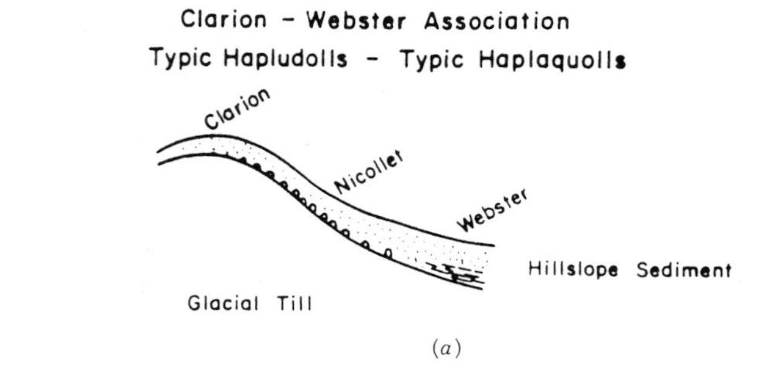

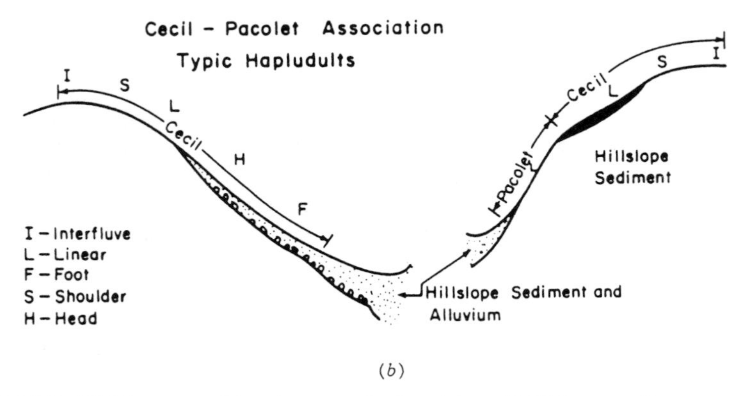

FIGURE 5.3. Thickness and landscape relationships of hillslope sediments and soils. A. Clarion-Webster association in Iowa and B. Cecil-Pacolet association in the North Carolina Piedmont. A. generalized and redrawn from Walker, 1966, Figs. 19 and 22, pp. 859, 860.

abrupt increase in mean diameter at that point. Below the sand lens, the mean diameter again decreased with distance downslope (Fig. 5.4 A and B).

Table 5.1 is data from a small watershed in southwestern Iowa and shows the changes in particle size of hillslope sediment related to its upslope source. Profile N's watershed has Tazewell loess and Kansan glacial till. The clay and silt content of the hillslope sediment (I horizons) changes when the sub-watershed is Late Sangamon paleosols and Kansan till. Organic carbon, exchange capacity and extractable Ca and Mg change with source area. Riecken and Poetsch (1960) reported similar relationships.

Bulk density, percentage clay and thickness of hillslope sediment change with distance of transport, or distance from the local summit (Fig. 5.5). Data from Malo et al. (1974) and Walker (1966)—both closed systems—and Riecken and Poetsch (1960)—an open system—quantify these progressive changes in sediment properties from the summit. Similar changes are easy to see in the field, and more important, are predictable.

Hillslope sediment properties, such as thickness or texture, and their relationships to the landscape are not random. Prediction value of this material is high if one considers the source area and shape of the landscape. Hillslope sediments are responsible for much, but not all, of the textural and yield variability found within a sloping field. Recent work (Ciha, 1984; Day et al., 1987; Jones et al., 1989; Schimel et al., 1985; Spratt & McIver, 1972) recognizes differences in crop yields and soil test values by landscape position. The strong correlation between landscape position, hillslope sediments and subsurface hydrology probably influenced their results.

Butler et al. (1986) studied the sequence of soils and vegetation in the Little Missouri Badlands of North Dakota. They found that vegetative cover varied with position in the hillslope and aspect. Clay content of soil increased downslope from the summit. Soil moisture deficits decreased from the back slope to the foot slope to the toe slope. The authors did not identify hillslope sediments, but the changes in texture and other properties suggest they occur in the area.

Unpublished data by the senior author show that landscape position has considerable influence on grain yields of corn (Fig. 5.6). Grain yields from

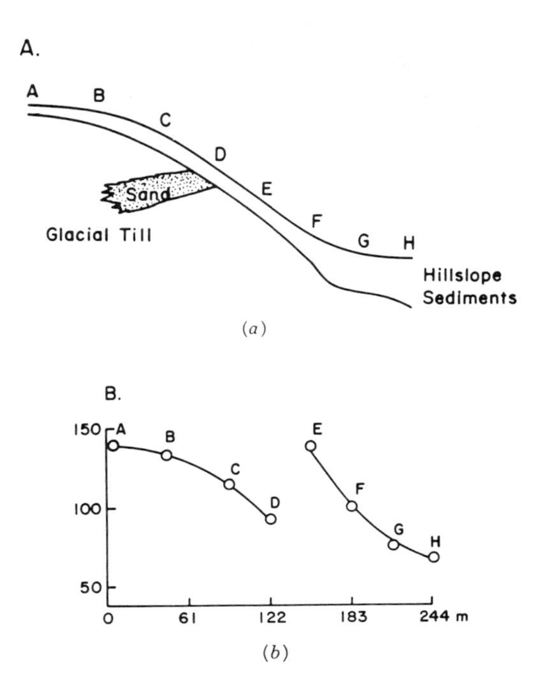

FIGURE 5.4. Changes in hillslope sediment downslope. A. Cross-section and sampling location. B. Weighted geometric mean diameter of hillslope sediment. Redrawn from Kleiss, 1970. A: Fig. 1, p. 288. B: Fig. 3, p. 289. Republished by permission from H.J. Kleiss.

TABLE 5.1. Some Physical and Chemical Properties of Soils Formed Partly in Hillslope Sediment and Partly in Glacial Till

Depth (in inches)	Horizon	Silt (%)	Clay (%)	>2mm (%)	Ratio silt/clay	Organic carbon (%)	CEC	Exch. Cat. Ca (meq/100g)	Exch. Cat. Mg 1 (meq/100g)
				Profile N *Hillslope Source Tazewell Loess and Till*					
0–5	IA11	57.0	35.6	<1	7.7	2.53	32.5	16.8	5.2
5–15	IA12	52.1	36.3	<1	4.5	1.81	31.2	16.0	4.8
15–19	IAB	46.9	34.5	<1	2.5	1.23	27.5	14.8	4.6
19–22	IB21	42.5	33.8	3	1.8	0.97	22.7	12.1	3.5
22–24	IB22	37.4	32.8	35	1.2	0.63	22.4	12.6	3.8
24–34	IIB23	34.2	33.4	1	1.0	—	21.6	13.9	4.3
34–45	IIB21	35.2	33.8	1	1.1	—	20.1	12.7	3.1
45–55	IIB32	38.5	29.5	4	1.2	—	Calcareous		—
55+	IIC	38.2	22.1	6	1.0	—	Calcareous		—
				Profile O *Hillslope Source Late Sangamon Paleosol and Till*					
0–6	IAp	34.9	25.8	<1	0.9	1.77	22.1	10.0	2.8
6–12	IA12	35.9	27.0	<1	1.0	1.41	21.7	9.9	2.8
12–16	IAB	36.7	28.0	1	1.0	0.97	20.3	10.3	2.8
16–21	IB21	36.6	27.0	3	1.0	0.70	18.7	10.5	2.8
21–23	IB22	33.4	25.4	23	0.8	0.51	16.9	10.2	2.1
23–30	IIB23	32.8	24.3	3	0.8	—	16.5	10.0	2.7
30–40	IIB31	32.4	24.6	2	0.8	—	14.4	10.2	1.4
40–50	IIB32	34.8	31.2	<1	1.0	—	18.1	13.2	2.4
50–61	IIC	37.6	27.9	4	1.0	—	Calcareous		—

From Ruhe et al., 1967, Table 36, p. 193.

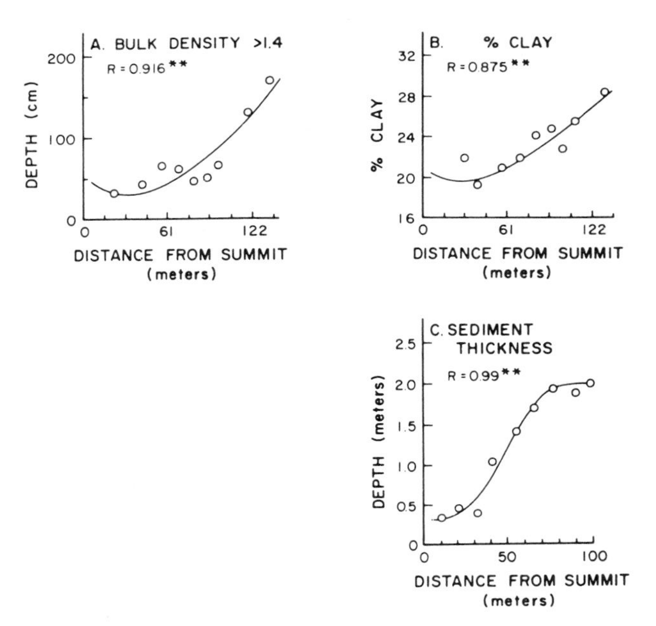

FIGURE 5.5. Changes in properties of hillslope sediment with increasing distance from the adjacent summit. A. Bulk density. B. Percent clay. C. Sediment thickness. Redrawn from Malo et al., 1974, Fig. 5, p. 816, Fig. 10, p. 817 and Walker, 1966, Fig. 21, p. 860.

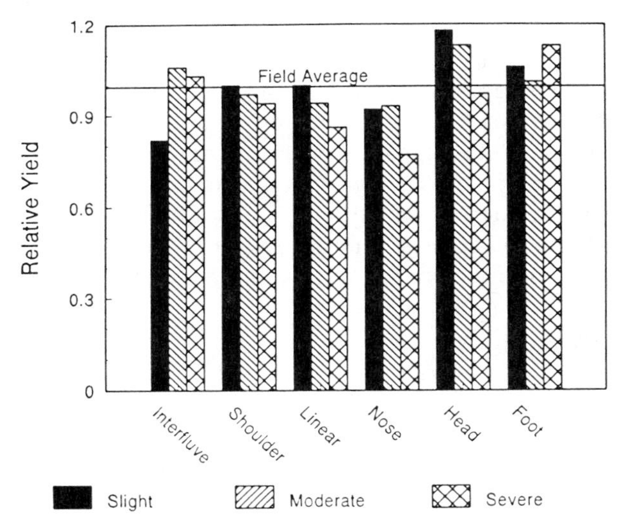

FIGURE 5.6. Relative yield of corn (*Zea Mays*) by landscape position and erosion class. (All data are on an area basis. Field production = 100.) Unpublished data from senior author.

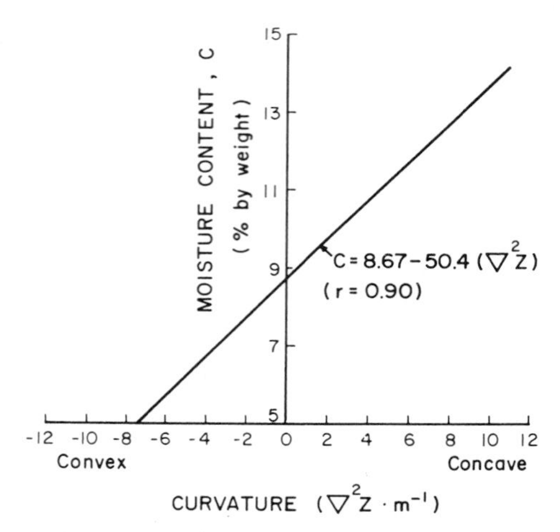

FIGURE 5.7. Changes in moisture content after a rain as influenced by slope curvature. Convex areas have negative curvature values and concave areas have positive values. Actual data points are not shown, but the regression line clearly shows more moisture after a rain in the concave than in the convex areas. Modified from Sinai et al., 1981, Fig. 9, p. 373. Reprinted by permission from Williams and Wilkins and G. Sinai.

corn are highest in areas where hillslope sediment is the thickest, the head and foot slopes, despite erosion class. These positions are also sites that receive moisture from the higher areas (O'Loughlin, 1981; Sinai et al., 1981).

The moisture content of the soil can vary considerably from the concave head and foot slope sites to the adjacent nose or linear slopes (Fig. 5.7). Neither Sinai et al. (1981) nor O'Loughlin (1981) recognized hillslope sediments, but landscape shape and position suggest these materials influenced their results. The downslope transfer of moisture is partly the result of the anisotropic nature of soils and hillslope sediments. In both natural and agronomic systems, the transfer of moisture has a large impact on the biology of the site.

REFERENCES

Bettenay, E. and F.J. Hingston. (1964). *Australian J. Soil Res.*, 2:173–186.

Burras, C.L. and W.H. Scholtes. (1987). *Soil Sci. Soc. Amer. J.*, 51:1541–1547.

Butler, B.E. (1958). *CSIRO Australian Soil Publ.*, no. 10.

Butler, B.E. (1959). *CSIRO Australian Soil Publ.*, no. 14.

Butler, J., H. Goetz, and J.L. Richardson. (1986). *Amer. Midland Naturalist*, 116:378–386.

Churchward, H.M. (1961). *J. Soil Sci.*, 12:73–86.

Churchward, H.M. (1963). *Australian J. Soil Res.*, 1:117–128.

Ciha, A.J. (1984). *Agron. J.*, 76:193–196.

Day, L.D., M.E. Collins, and N.E. Washer. (1987). *Soil Sci. Soc. Amer. J.*, 51:1547–1553.

Eargle, D.H. (1940). *Science*, 91:337–388.

Hack, J.T. and J.C. Goodlett. (1960). *Geomorphology and Forest Ecology of a Mountain Region in the Central Appalachians*. U.S. Geol. Survey Prof. Paper, No. 347.

Huddleston, J.H. and F.F. Riecken. (1973). *Soil Sci. Soc. Amer. Proc.*, 37:264–270.

Jessup, R.W. (1961). *J. Soil Sci.*, 12:52–63.

Jones, A.J., L.N. Mielke, C.A. Barties, and C.A. Miller. (1989). *J. Soil and Water Conserv.*, 44:328–332.

Kleiss, H.J. (1970). *Soil Sci. Soc. Am. Proc.*, 34:287–290.

Malo, D.D., B.K. Worcester, D.K. Cassel, and K.D. Matzdorf. (1974). *Soil Sci. Soc. Am. Proc.*, 38:813–818.

Matzdorf, K.D., D.K. Cassel, B.K. Worcester, and D.D. Malo. (1975). *Soil Sci. Soc. Am. Proc.*, 39:508–512.

McCracken, R.J., R.B. Daniels, and W.E. Fulcher. (1987). *Soil Sci. Soc. North Carolina Proc.*, 30:88–108.

Milne, G. (1936). *Nature*, 138:548–549.

O'Loughlin, E.M. (1981). *J. Hydrology*, 53:229–246.

Riecken, F.F. and E. Poetsch. (1960). *Iowa Acad. Sci.*, 67:268–276.

Robinson, G.W. (1936). *Nature*, 137:950.

Ruhe, R.V. (1959). *Soil Sci.*, 87:223–231.

Ruhe, R.V., R.B. Daniels, and J.G. Cady. (1967). *Landscape Evolution and Soil Formation in Southwestern Iowa*. USDA Tech. Bull. 1349.

Schimel, D., M.A. Stillwell, and R.G. Woodmansee. (1985). *Ecology*, 66:276–282.

Sinai, G., D. Zaslavsky, and P. Golany. (1981). *Soil Sci.*, 132:367–375.

Soil Survey Staff. (1975). *Soil Taxonomy*. [Agric. Handbook 436] Washington, DC: U.S. Govt. Printing Office.

Spratt, E.D. and R.N. McIver. (1972). *Can. J. Soil Sci.*, 52:53–58.

Van Dijk, D.C. (1959). *CSIRO Australian Soil Publ.*, no. 13.

Vreeken, W.J. (1988). *Iowa Acad. Sci. Proc.*, 75:225–233.

Walker, P.H. (1962a). *J. Soil Sci.*, 13:167–177.

Walker, P.H. (1962b). *J. Soil Sci.*, 13:178–186.

Walker, P.H. (1966). Postglacial environment in relation to landscape and soils on the Cary Drift, Iowa. *Iowa Agri. Exp. Sta. Res. Bull.*, 549:838–875.

Whittecar, G.R. (1985). *SE Geol.*, 26:117–129.

6 Transitional Environments and Terrigenous Marine Shelf Sediments

INTRODUCTION

Deltas, estuaries and barrier systems are the major types of transitional environments between continental and marine deposits. Sediments of deltas and estuaries usually form elongated bodies oriented at right angles to the paleo-shore line, whereas barrier and associated facies parallel the shoreline. The shoreface and terrigenous marine shelf sediments are also complex units, but they may be of greater areal extent than the transitional sediments.

The most intensively cultivated areas throughout the world are transitional sediments, deltas, barrier systems and uplifted marine sediments. In both deltas and coastal plains, large areas retain their depositional shape, so predictions are good of soil properties and textures inferred from landform. Transitional and marine environments are widespread throughout the geologic column, and exposure of these sediments is common in erosional landscapes. It is difficult to explain or partly understand the variability of soil materials unless one understands the sediment's depositional environment.

DELTAS

Deltas develop wherever a river or stream carrying large quantities of sediment discharges into a water basin (Elliott, 1975; Galloway, 1975). The basins can be fresh or saline lakes, estuaries, lagoons or the ocean. Deltas are essentially alluvial fans (subaerial units) deposited subaqueously beneath either fresh or salt water. By contrast, streams deposit fans subaerially. Sediment carried by channel flow discharging into a reservoir spreads and mixes with basin water. Discharge velocity and density contrast of river and basin water determine the rate and geometry of sediment spreading and mixing (Fig. 6.1; Bates, 1953; Elliott, 1975; Fisher et al., 1969). River water is more dense than fresh basin water and flow; mixing and deposition occurs at the fresh water base. In marine basins, the salt water is more dense than the river water, and flow and mixing take place near the surface. Similar differences in the mixing zone occur in lakes when water temperatures differ between the influent and the basin.

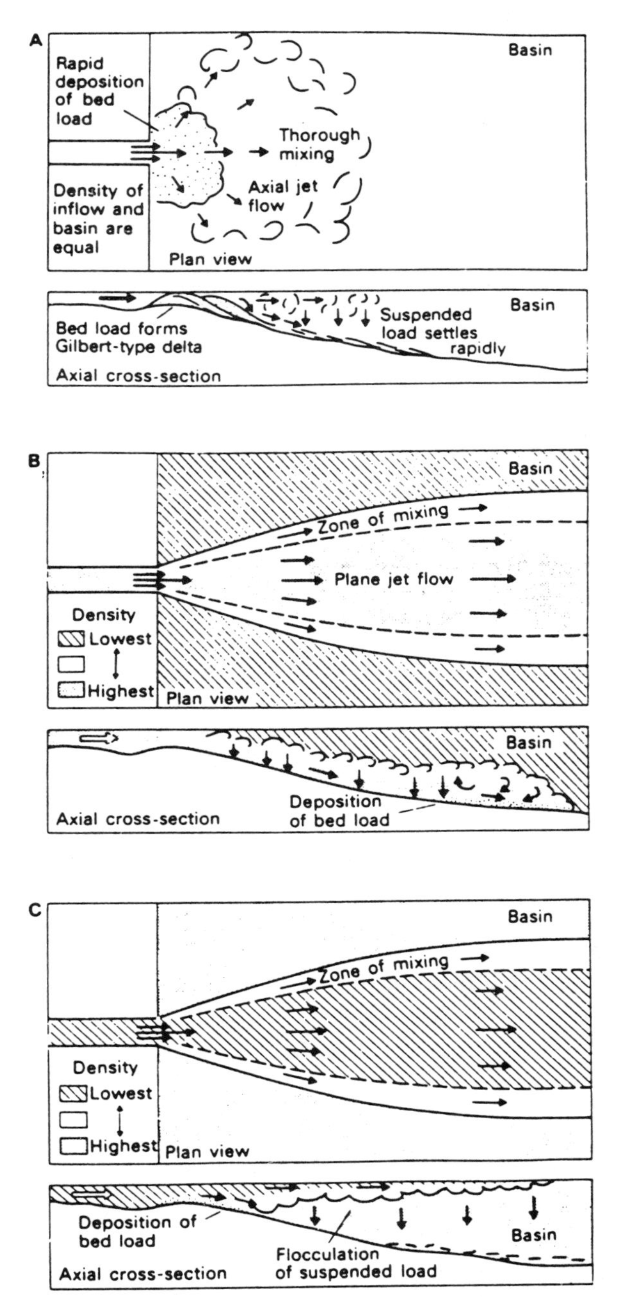

FIGURE 6.1. Deposition of bedload and location of mixing zones where (a) waters are of equal density, (b) influent is more dense than basin water, and (c) influent less dense than basin water. From Fisher et al., 1969, Figs. 2, 3 and 4 and Bates, 1953. Republished by permission from W.L. Fisher.

River Dominated Deltas

Several depositional environments occur near the delta channel mouth (Fig. 6.2; Coleman & Gagliano, 1964; Fisher et al., 1969; Galloway & Hobday, 1983; Wright, 1985), but the sediments fine seaward from the bar to the prodelta (Galloway & Hobday, 1983; Wright, 1985). Major floods carry bedload onto the prodelta slope and deposit it as aprons, lobes of sand, or frontal splays and thus interrupt the fining seaward (Fig. 6.2a).

Flood waters also breach levees of the lower delta and form crevasse splays that become subdeltas prograding onto the adjacent interdeltaic embayments (Fig. 6.3; Coleman & Gagliano, 1964). Splays fill the embayments largely during flood periods, and are wedges that thicken away from the apex. The coarsest sediments are at the apex, and particle size decreases toward the distal edge. The crevasse splays along delta distributaries are important in developing the delta plain. The splays develop rapidly (Welder, 1959) but sporadically, as floods are the major source of sediments.

Channels in the lower delta are not stable, and channel shifting (avulsion) to courses of higher slope is common. In the Mississippi delta, 16 such lobes have developed from major channel changes during the last 6,000 years. Through Holocene, the locus of deposition moved as much as 240 km along the basin margin (Frazier, 1967).

Wave energy and longshore currents (Fig. 6.2b) mobilize the channel sands and transport them in the direction of the longshore drift (Fisher et al., 1969). The sand belt is broad if the sand stays in place. Several coalescing arcuate beach ridges form where longshore currents rework the sand. These sand ridges concentrate the coarsest material available, including shells. Eolian processes play an important role in sorting and distributing the coastal sands (Goldsmith, 1985). Shallow bays and extensive salt marshes develop between the mainland and the beach ridges. Bay mud, shells and silts form in the deeper bays; peat occupies the shallow parts (Fig. 6.4; Fisher et al., 1969). Sediments of prograding deltas coarsen upward as the environment changes from prodelta through interdistributary environments (Fig. 6.2; Fig. 6.5; Coleman & Wright, 1975).

Wave- and Tide-Dominated Deltas

Wave-dominated deltas are similar to fluvial dominated systems with one major exception. Waves and longshore currents rework sediments deposited at the mouth and widely disperse suspended sediments. The abundant coastal

FIGURE 6.2. A. Channel mouth depositional environments. From Wright and Coleman, 1974, Fig. 14, p. 778. Reprinted by permission from University of Chicago Press. B. Channel mouth processes. Modified from Galloway and Hobday, Fig 5-2B, p. 83. Reprinted by permission from Springer-Verlag and W.E. Galloway. Originally modified from Fisher et al., 1969.

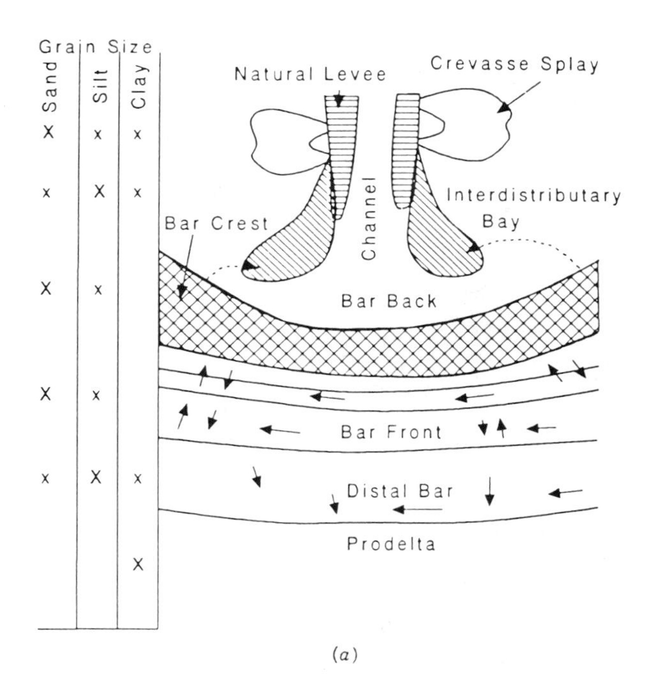
Grain Size
Sand
Silt
Clay
Natural Levee
Crevasse Splay
Channel
Interdistributary Bay
Bar Crest
Bar Back
Bar Front
Distal Bar
Prodelta

(a)

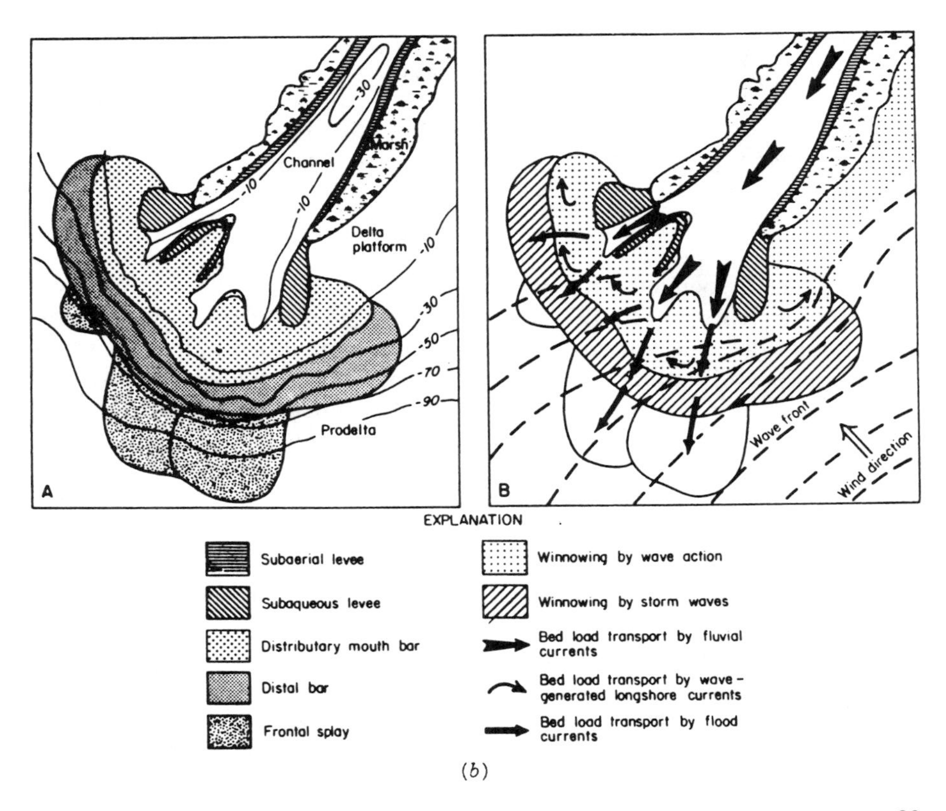
Channel
Marsh
Delta platform
-30
-10
-50
-70
-90
Prodelta
A
B
Wave front
Wind direction
EXPLANATION
Subaerial levee
Subaqueous levee
Distributary mouth bar
Distal bar
Frontal splay
Winnowing by wave action
Winnowing by storm waves
Bed load transport by fluvial currents
Bed load transport by wave-generated longshore currents
Bed load transport by flood currents

(b)

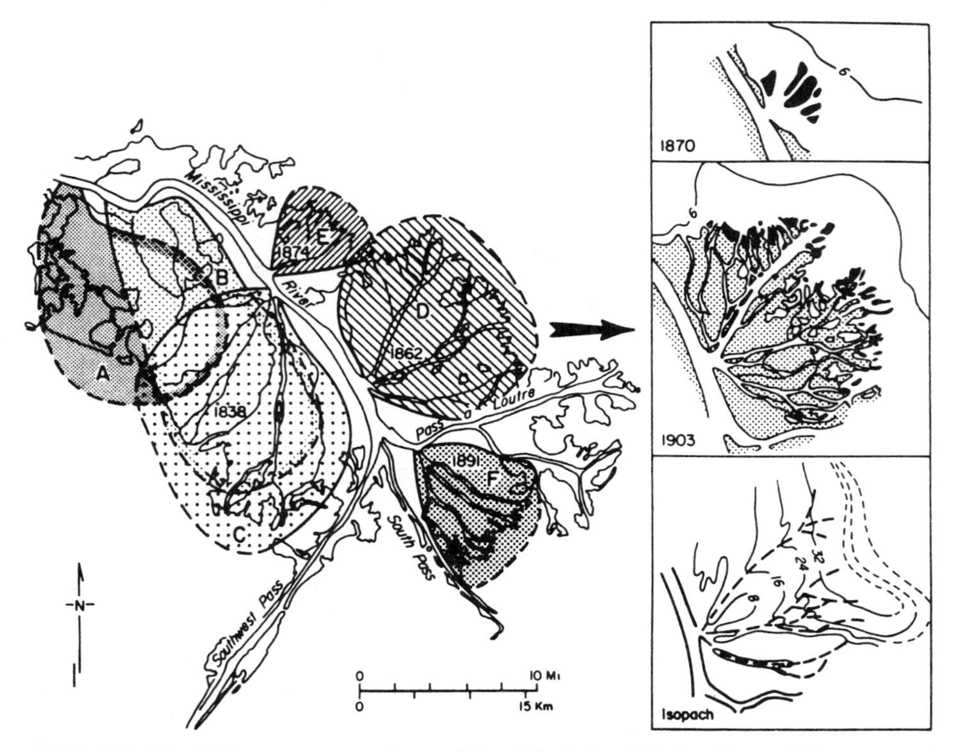

FIGURE 6.3. Major crevasse splays of the Mississippi Delta. From Coleman and Gagliano, 1964, Fig. 5, p. 75. Reprinted by permission from Gulf Coast Association of Geological Societies.

beach ridges associated with wave-dominated deltas result in prominent coastal dune fields (Elliott, 1975; Galloway & Hobday, 1983). Sediment transport in tidal deltas is primarily onto the prodelta platform (Fig 6.2a) where rapid mixing and settling of suspended sediment occurs. Figures 6.6 and 6.7 are a cross-section and idealized sequence of sediments in a tide-dominated delta. Table 6.1 gives the characteristics of deltaic depositional systems.

ESTUARIES

Drowned river mouths (estuaries) usually occur where stream sediment load is <160 mg/l, but geologists frequently disagree on what constitutes an estuary (Cronin, 1975; Lauff, 1967; Roy et al., 1980). By contrast, deltas develop where river sediment loads are >225 mg/l (Friedman & Sanders, 1978).

Estuarine facies include channel muds, silt and sands passing seaward into tidal deposits and landward into river deposits (Nichols & Biggs, 1985).

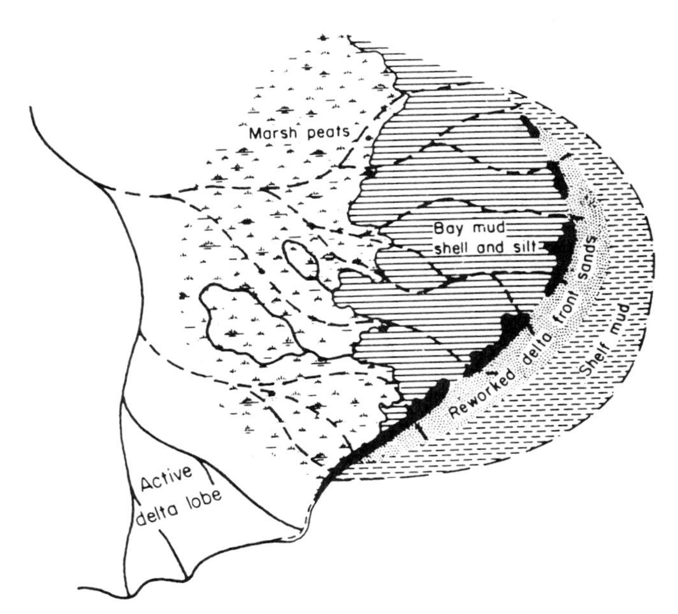

FIGURE 6.4. Environments produced by reworking of an abandoned fluvial-dominated delta lobe. From Galloway and Hobday, 1983, Fig. 5-13, p. 99; based on Fisher et al., 1969, Fig. 24. Reprinted by permission from Springer-Verlag and W.E. Galloway.

TABLE 6.1. Characteristics of Deltaic Depositional Systems

	Domination		
	Fluvial	Wave	Tide
Geometry	Elongate to lobate	Arcuate	Estuarine to irregular
Composition	Muddy to mixed	Sandy	Muddy to sandy
Facies	Distributary bar and delta front sheet sand; distributary channel fill sand	Coastal barrier sand; distributary sand	Tidal sand ridge sand; estuarine distributary channel fill sand
Orientation	Variable; parallels dep. slope	Parallels dep. strike; subsidiary dip trends	Parallels depositional slope unless skewed by local basin geometry
Channel type	Suspended-load to fine mixed-load	Mixed-load to bedload	Variable, tidally modified geometry

Modified from Galloway (1975), Table 2, p. 89. Reprinted by permission from W.E. Galloway.

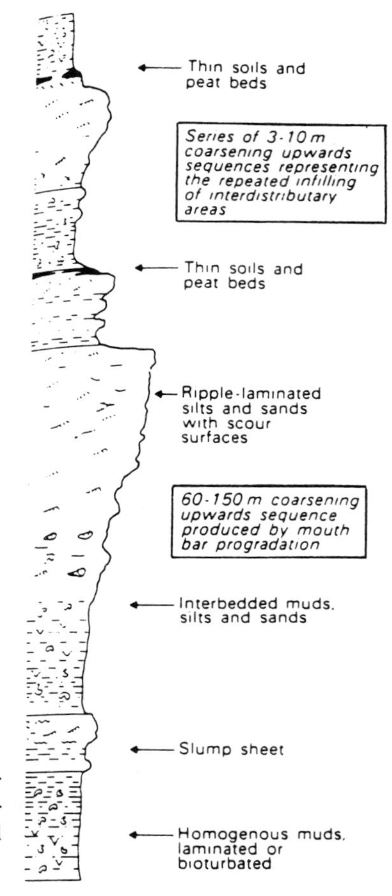

FIGURE 6.5. Upward coarsening sediments produced by mouth bar progradation. After Coleman and Wright, 1975, Fig. 13, p. 129. Reprinted by permission from J.M. Coleman.

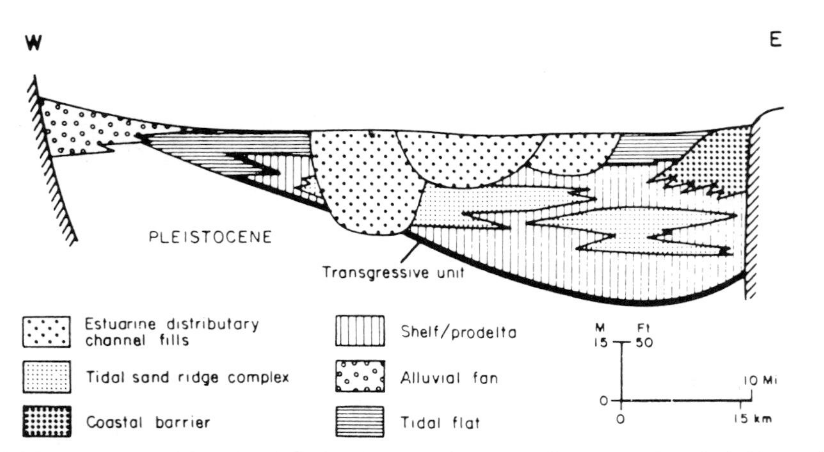

FIGURE 6.6. Depositional units of a tide-dominated delta. From Meckel, 1975, Fig. 18, p. 263. Reprinted by permission from L.D. Meckel.

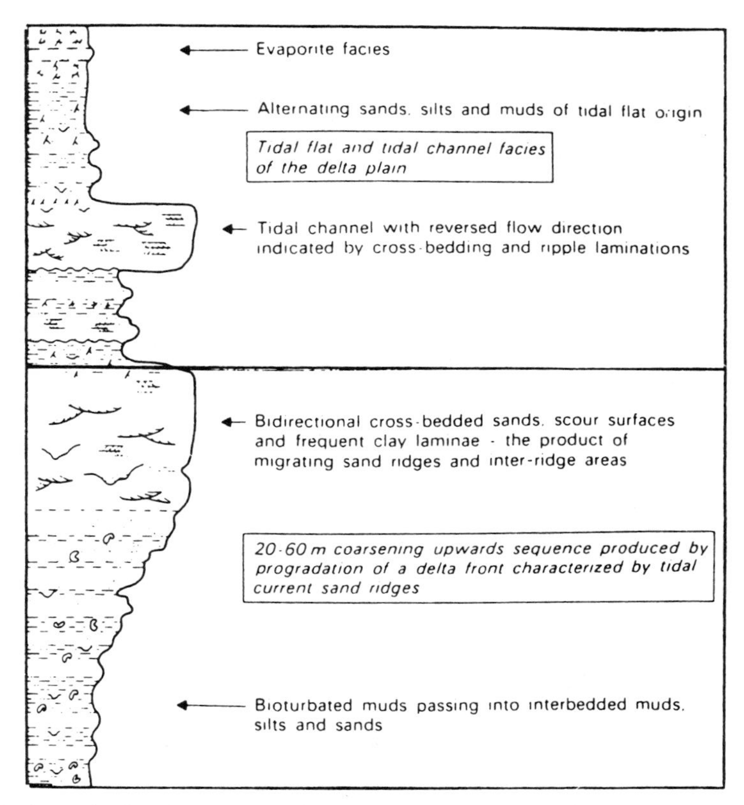

FIGURE 6.7. Idealized sequence of sediment through a tide-dominated delta. After Coleman and Wright, 1975, Fig. 19, p. 136. Reprinted by permission from J.M. Coleman.

Fluvial sands near the heads of estuaries merge downstream with muds. Estuaries also fine upstream from the sandy seaward portions to mud accumulation at the turbidity maximum. Figure 6.8 is a cross-section of the upper Chesapeake Bay showing the relationship among sediments and position in the estuary. Figure 6.9 shows the dispersal zones and the major processes within an estuarine system.

Ancient estuarine sediments are difficult to identify, and major proof of this environment is its gross shape and the presence of salinity-sensitive fauna and flora. Estuaries are geologically ephemeral features. Sediment fills estuaries from seaward by washover sands and landward by bayhead deltas (Friedman & Sanders, 1978). The estuarine environment merges horizontally seaward with the complex environments of the barrier/lagoon system.

BARRIERS AND ASSOCIATED ENVIRONMENTS

Barriers and associated environments are part of a delta system and an independent shore-zone system (Galloway & Hobday, 1983). The barriers, beaches,

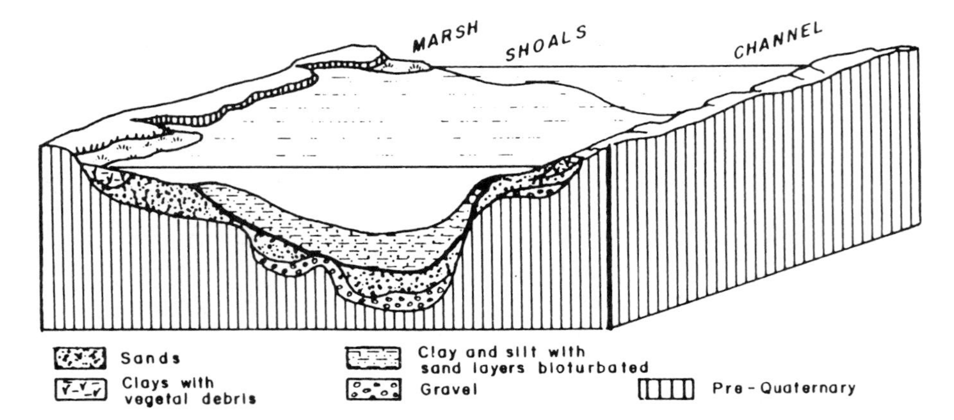

FIGURE 6.8. Relationship of upper Chesapeake Bay sediments to position in the estuary. From Nichols and Biggs, 1985. Fig. 2-11, p. 93. Reprinted by permission from Springer-Verlag, M.M. Nichols and R.B. Biggs.

lagoons and tidal flats are the result of marine reworking of sediments as contrasted with the direct fluvial origin of deltas. Sands dominate barrier systems, but tidal flats and lagoons have fine-textured materials. The shore zone and associated features have complex depositional environments that change rapidly at right angles to the shoreline (Figs. 6.10 and 6.11). These shore-zone facies usually produce linear bodies parallel to the shore line and result in abrupt vertical and horizontal changes in sediments.

Lagoon Deposits

The lagoon and bay, barrier and inlets are major elements of the barrier/lagoon system. The lagoon has many estuarine environments, but tides and nearshore marine processes dominate. Lagoons and associated bays have sediments from bayhead deltas, washover sands and inlet tidal deltas.

The common coastal marshes in lagoon systems are sediment sinks. Lagoons receive sediments from land, washover fans and landward transport of shelf sediments through inlets. Marsh sediments have sand to clay textures, and position in the marsh controls grain size (Figs. 6.12 and 6.13; Edwards & Frey, 1977). Tidal flats are low relief, unvegetated areas that can be either mud or sand. Gravel is uncommon in tidal flats (Davis, 1983). Within the tidal flat, the sediments are coarse near low-tide and fine near high-tide levels (Fig. 6.14).

Muds accumulate in the lagoon by flocculation and biogenic pelletization of suspended clays (Galloway & Hobday, 1983). Coarser sediments are at the margins, where they grade into tidal flats and marshes.

Washover fans form when storms erode a channel across the barrier and develop wedge-shaped sand deposits in the lagoon and backbarrier flat (Fig.

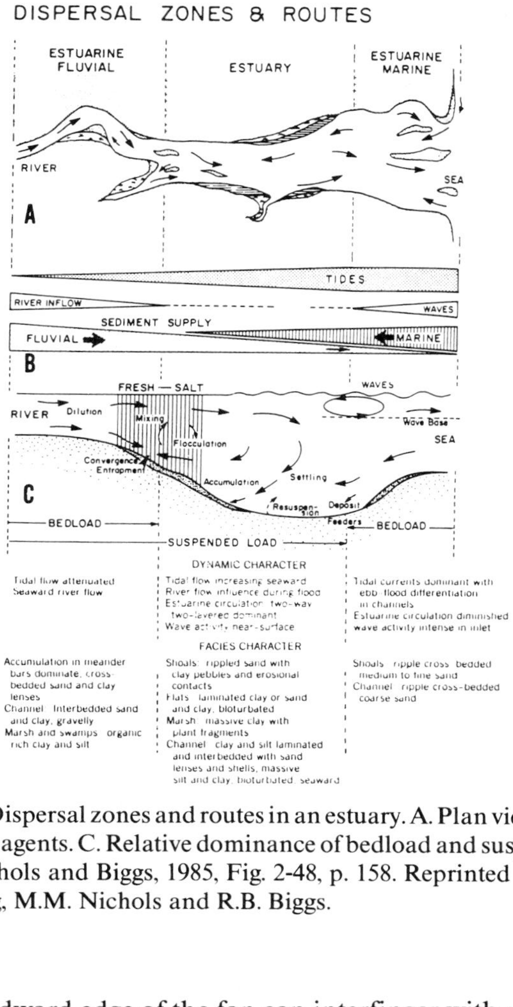

FIGURE 6.9. Dispersal zones and routes in an estuary. A. Plan view. B. Relative intensity of dispersal agents. C. Relative dominance of bedload and suspended load deposition. From Nichols and Biggs, 1985, Fig. 2-48, p. 158. Reprinted by permission from Springer-Verlag, M.M. Nichols and R.B. Biggs.

6.10). The landward edge of the fan can interfinger with marsh, tidal flat or lagoonal sediments. The basal sediments usually are shell-sand mixtures that grade upward into laminated sands. The sands are from the beach and barrier dunes.

Inlets exchange water between the backbarrier and the ocean. Closely spaced inlets have large tidal ranges and widely spaced inlets have low tidal ranges (Fig. 6.15; Barwis & Hayes, 1979; Elliott, 1975). Inlets are ephemeral features and migrate rapidly in the direction of longshore drift. Flood deltas develop into the lagoon or estuary and ebb deltas into the ocean.

Tidal delta sediments separate into several populations. A coarse lag deposit of shells, pebbles and other coarse particles develops on the frontal part of the delta (Boothroyd, 1985). Flood deltas transport the finer sands away from the channel and deposit them as the tidal currents lose competence. The

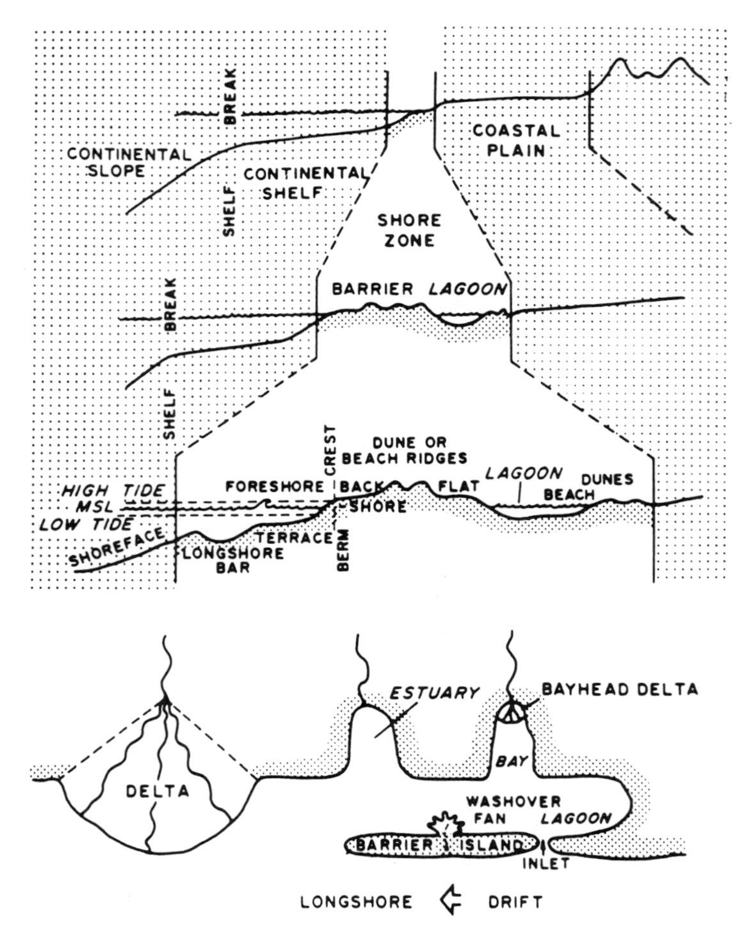

FIGURE 6.10. Schematic of shore-zone environments. A. Relationship to coastal plain and shelf. B. Morphological subdivisions. From Curray, 1969, Fig. JC II-1, p. JCII. Reprinted by permission from American Geological Institute.

inlets can transport large quantities of sediment into the lagoon. Sands of the ebb tidal deltas are subject to reworking by waves and longshore currents.

Barrier Deposits

Barriers are the offshore islands that separate the lagoon from the open ocean. Barriers have several distinct morphological shapes and are mostly sand. Nearly flat, low slope shelves and an abundant sediment supply, with low to moderate tidal ranges, favor barriers (Glaser, 1978). Authors suggest several origins for barriers, but many barriers are of composite genesis (Glaser, 1978;

Nummendal et al., 1977; Swift, 1975). Barrier systems include the shoreline features and the bay or estuary behind the barrier dunes (Fig. 6.16). The complex shoreline features range from dune to open fresh-water marsh environments (Fig. 6.17).

Tidal range controls barrier morphology (Fig. 6.15). Microtidal barriers are long and narrow with many washover fans and few inlets. Mesotidal barriers are broad with several inlets and tidal sand shoals. These barriers have a wide updrift end, a narrow midsection and a recurved downdrift end (Hayes, 1975). High-profile barriers (Fig. 6.18) occur along stable or prograding coasts, whereas low-profile barriers are long transgressive coasts (Morton & McGowen, 1980).

Figure 6.19 A and B shows the areal extent of barriers and associated sediments near Myrtle Beach, South Carolina (DuBar, 1971; Markewich et al., 1986). Thom et al. (1978) examine the different barrier morphologies and stratigraphic sequences associated with transgression and regression. Trans-

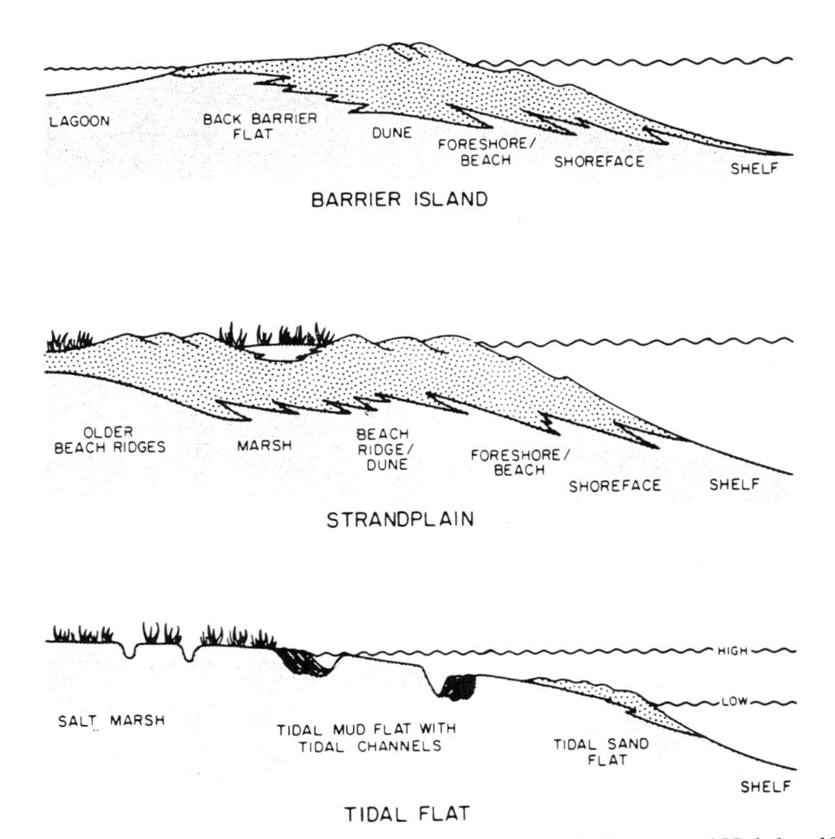

FIGURE 6.11. Contrasting coastal morphologies. From Galloway and Hobday, 1983, Fig. 6-2, p. 117. Reprinted by permission from Springer-Verlag and W.E. Galloway.

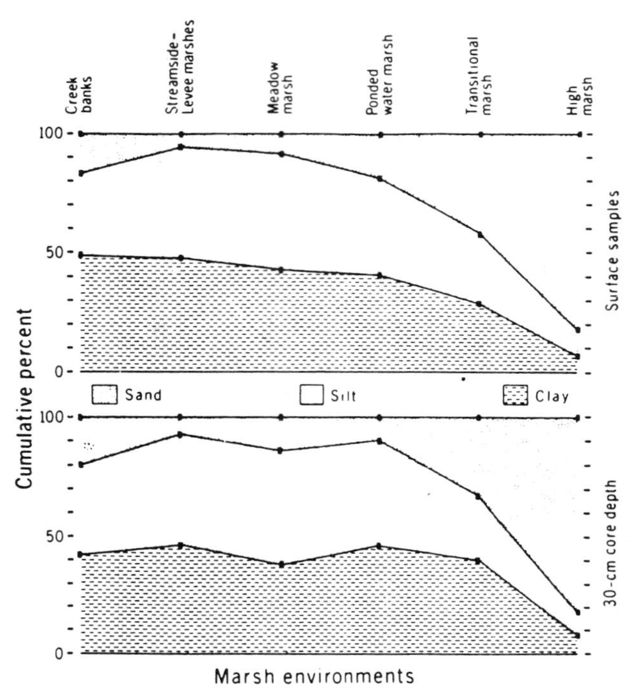

FIGURE 6.12. Textural properties of marsh sediments and relationship to microtopography. From Frey and Basan, 1985, Fig. 4-8, p. 247. Reprinted by permission from Senckenbergische Naturforschende.

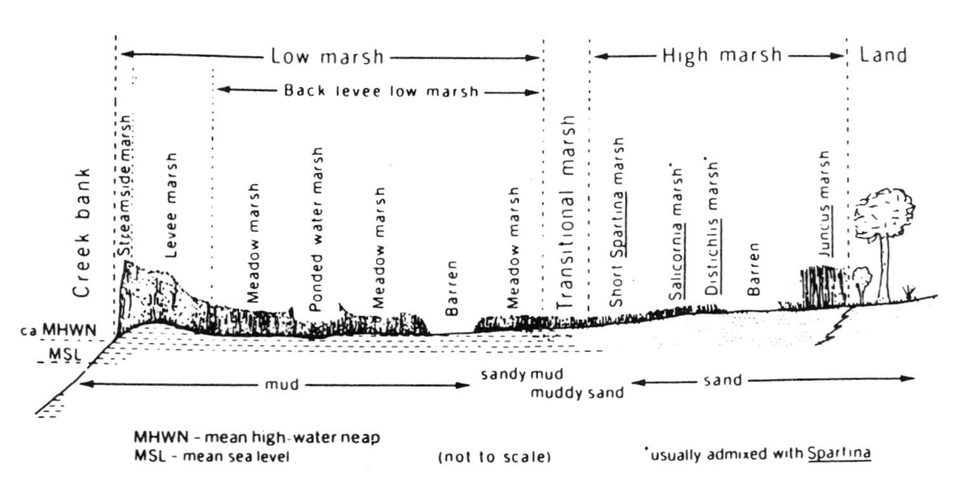

FIGURE 6.13. Relationship between sediment textures and high and low marsh in Georgia. From Edwards and Frey, 1977, Fig. 2, p. 219. Reprinted by permission from Senckenbergische Naturforschende.

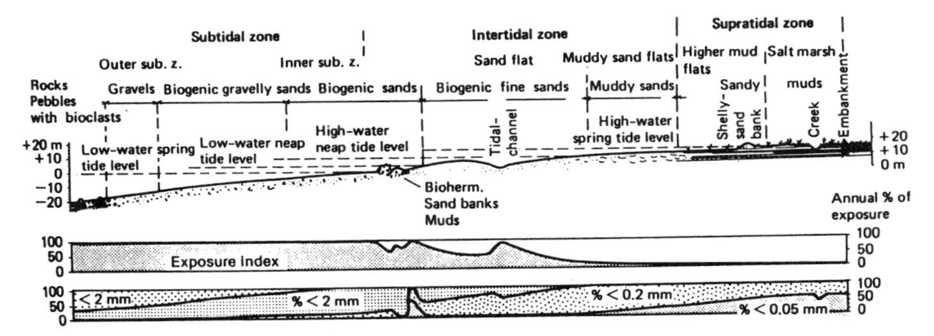

FIGURE 6.14. Zonation of tidal flats along the North Sea coast of Europe. Modified from Larsonnieur, 1975, Fig. 3.2, p. 24. Reprinted by permission from Springer-Verlag.

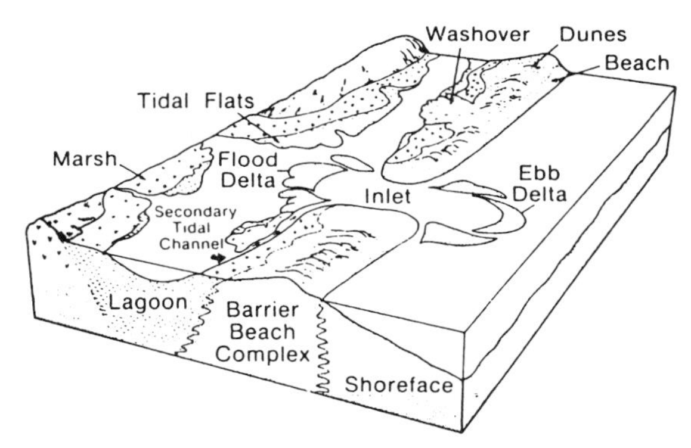

FIGURE 6.15. Depositional environments associated with barrier island shorelines. From Reinson, 1980, Fig. 2, p. 58. Reprinted by permission from Geological Association of Canada.

gressive and regressive barriers also have major differences in the areal extent, thickness and geographic distribution of associated facies (Figs. 6.20 and 6.21; Kraft, 1971; Kraft & John, 1979).

The major sedimentation processes of transgressive barriers are washover and inlet deposition. Flood-tidal deltas can spread landward large volumes of sand that serve as a platform for washover fans and eolian materials. Regressive barrier sediments coarsen upward from lower shoreface sand and clay to mixed shells and sand of the upper shoreface and beach (Fig. 6.21). Longshore migration of inlets obliterates much of the barrier-beach and shoreface sediment (Davis, 1983; Galloway & Hobday, 1983). The inlet-fill deposits replace these sediments and may be as much as three times the thickness of the original barrier sands (Hoyt & Henry, 1967). The inlet-fill deposits in microtidal inlets are 6 to 12 m thick, and in mesotidal inlets may be 20 m thick. The

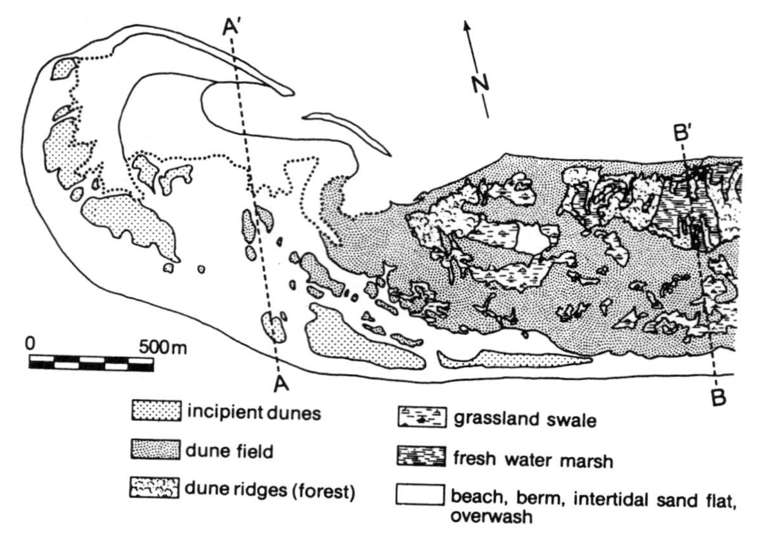

FIGURE 6.16. Geomorphic features of a barrier island. From Moslow and Heron, 1989, Fig. 31a, p. T171:16. Reprinted by permission from American Geophysical Union.

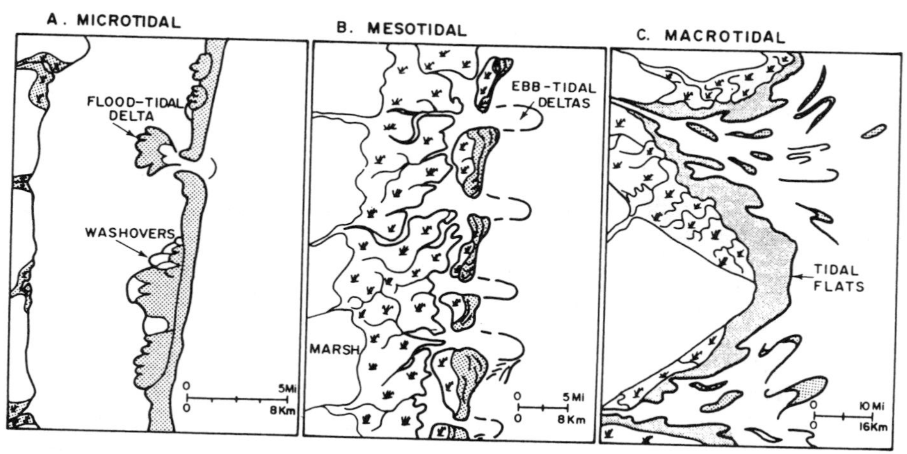

FIGURE 6.17. Tidal controlled morphology of barriers. From Barwis and Hayes, 1979, Fig. 2, p. 473. Reprinted by permission from J.C. Ferm.

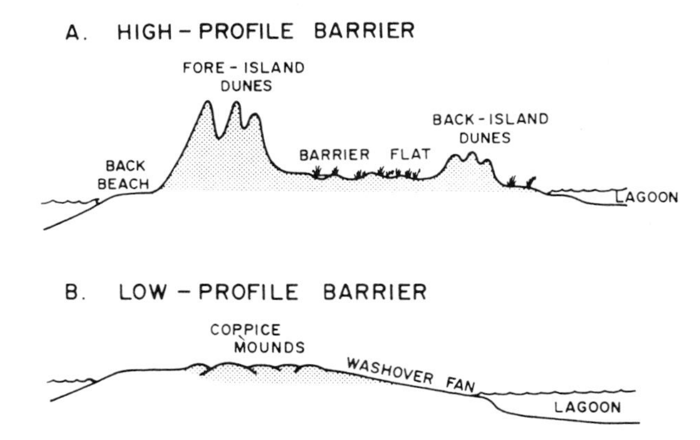

FIGURE 6.18. High- and low-profile barriers with associated features. From Galloway and Hobday, 1983, Fig. 6-9, p. 125. Reprinted by permission from Springer-Verlag and W.E. Galloway.

geometry and sequence of sediments in either prograding or transgressing systems are very complex (Barwis & Hayes, 1979).

TRANSITIONAL SYSTEMS AND SOILS

Barrier systems are widespread throughout coastal areas and form the parent materials for important agricultural and forested soils. The barrier usually forms offshore, and a lagoon or backbarrier flat develops between the barrier and the mainland (Fig. 6.21). Distinctive suites of soils of similar age develop on the barrier and the backbarrier flats.

Some barriers form on the mainland (Fig. 6.22), so only the soils seaward and those in the equivalent estuaries or lagoons are the same age as the barrier. The soils behind the barrier on the Talbot Plain of Figure 6.22 are older than the soils on the Arapahoe Ridge or the Pamlico Plain. The Arapahoe Ridge and its association with the Talbot Plain is an example of why the stratigraphic relations must be proven. Form alone often leads to incorrect conclusions.

Barrier systems may be wide, such as those in South Carolina (Fig. 6.17), or very narrow as the Arapahoe Ridge in North Carolina (Fig. 6.22 A and B). These systems may be smooth, broad flats with little relief, or irregular as is the prograding beach ridge system south of the Neuse River in North Carolina (Fig. 6.23). Most barrier systems are a series of barrier and backbarrier flats.

Soils formed in the barrier sediments typically are 80 to 95 percent sand, with fine and medium sands dominating (Markewich et al., 1986, Fig. 3, pp. 8–9). Spodic or humate horizons are common in the sandy barriers (Fig. 6.24B; Daniels et al., 1976; Markewich et al., 1986). The spodic horizon may be a very

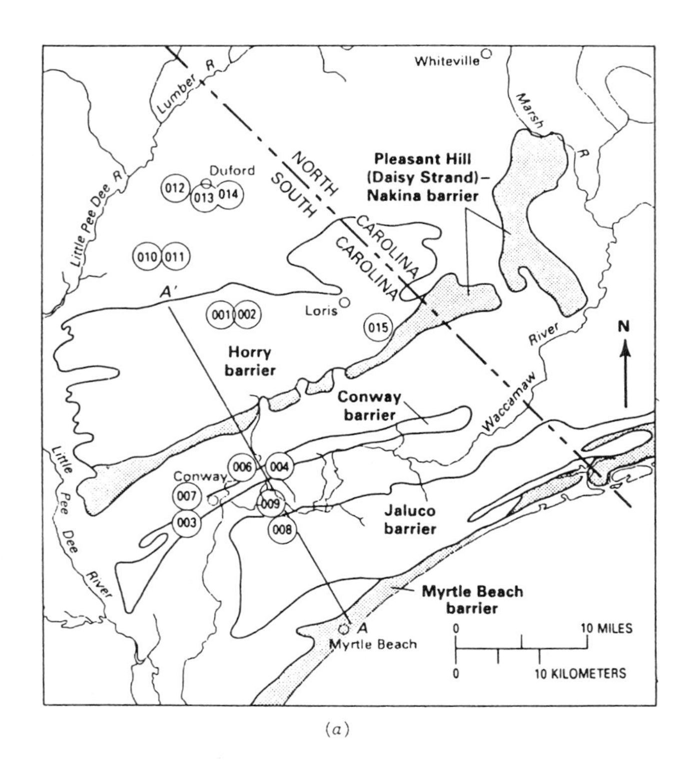

(a)

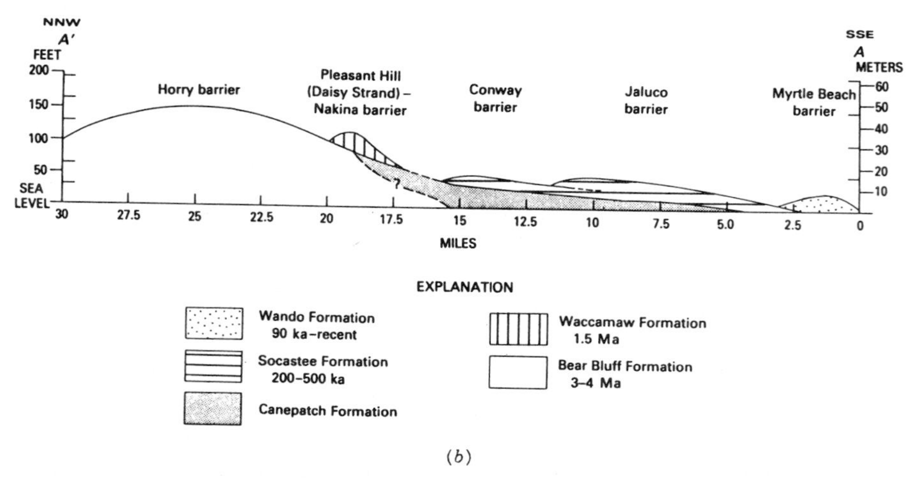

(b)

FIGURE 6.19. A. Areal extent of barriers and associated sediments near Myrtle Beach, South Carolina. From Markewich et al., 1986, Fig. 2A, p. 4; Modified from DuBar, 1971. B. Cross section of morphostratigraphic units in Horry County, South Carolina. From Markewich et al., 1986, Fig. 2B, p. 5. Reprinted by permission from H. Markewich and J. Dubar.

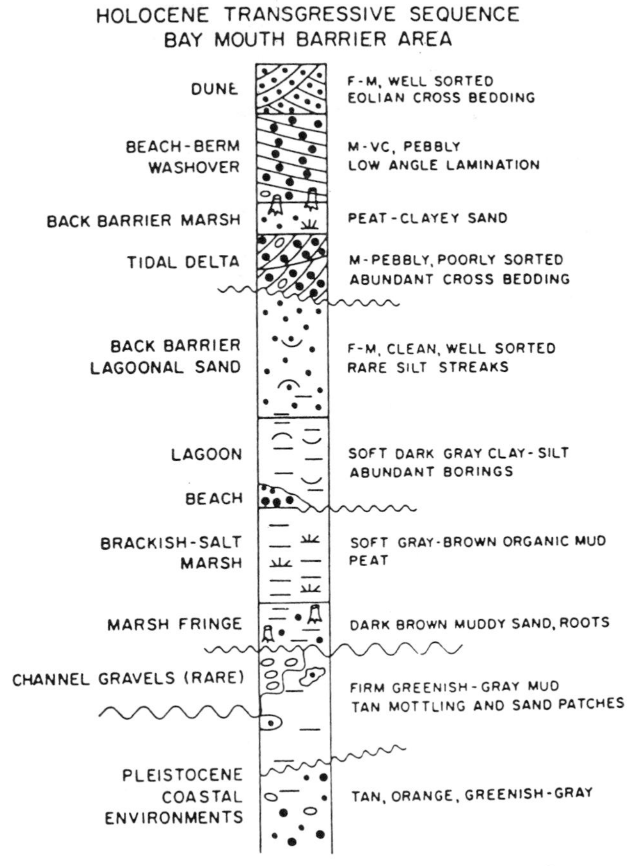

FIGURE 6.20. Transgressive sequence before erosion by shoreface processes. From Kraft, 1971, Fig. 23, p. 2153. Reprinted by permission from Geological Society of America and J.C. Kraft.

few centimeters (cm) thick beneath the dryer sites, or 6 to 7 meters thick beneath broad flats with deep water movement (Fig. 6.24 A and B; Daniels et al., 1975; Holzhey et al., 1975).

Most soils on the Arapahoe Ridge have spodic horizons (Fig. 6.22B), as do the soils on the beach ridge complex in Figure 6.23. The soils on the beach ridge complex are sands (Goodwin, 1987b). The A1 and spodic horizons on the ridges are often discontinuous and thin. The soils in the swales have thick organic-rich A horizons and distinct spodic horizons. Soil and vegetation change abruptly in response to differing hydrological gradients across the ridge and swale sequence.

Soils on backbarrier flats can have a wide range of textures. Many back-barrier systems are extensive, have a nearly level surface and have finer tex-

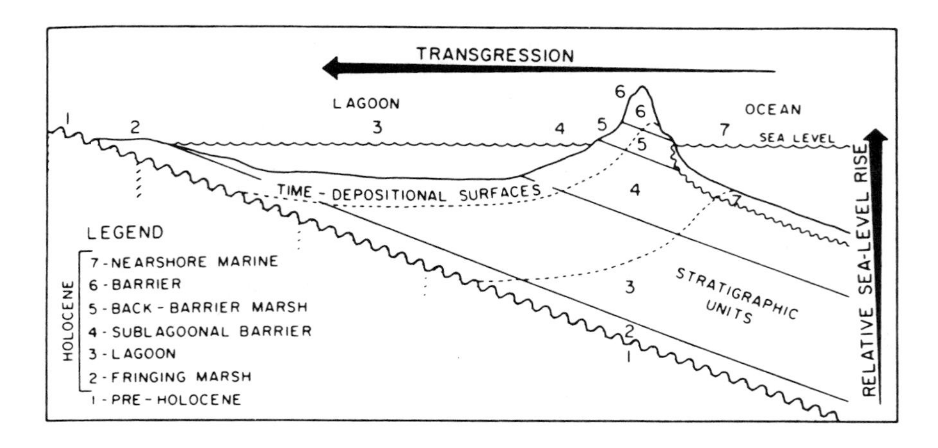

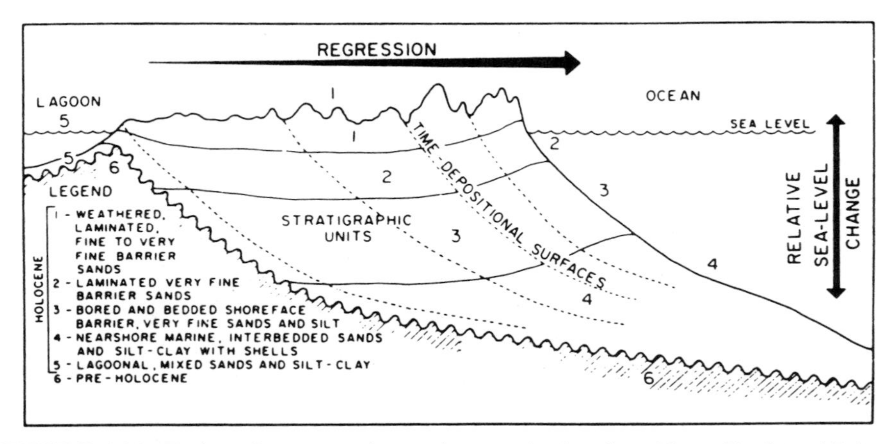

FIGURE 6.21. Facies of transgressive and regressive barriers. From Kraft and John, 1979, Fig. 12, p. 2161. Reprinted by permission from American Association of Petroleum Geologists.

tures than the adjacent barrier. Near Myrtle Beach the backbarrier soils are dominantly silty clay loams (Markewich et al., 1986, Fig. 8, pp. 14–15). Abundant organic soils over loamy to fine-textured sediments or mineral soils occur west of the Arapahoe Ridge (Fig. 6.22A). Fine-textured mineral soils also occur west of the ridge (Goodwin, 1987a).

SHOREFACE AND TERRIGENOUS MARINE SHELF SEDIMENTS

Sediments deposited in shallow seas on continental shelves are important soil materials in coastal plains. Most of these materials come from land. The marine deposits have a wide range of environments. Shelf deposits occur from

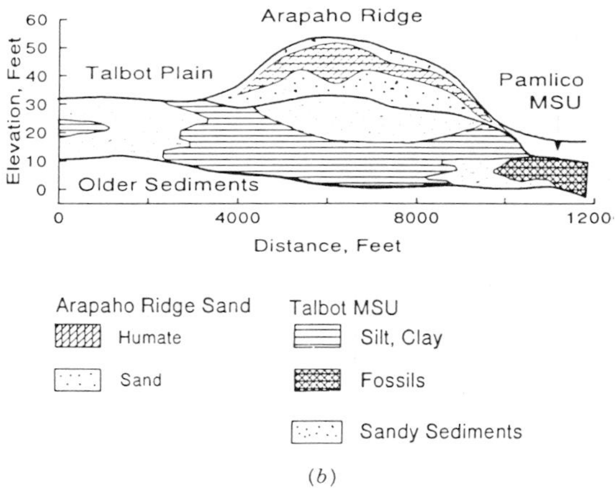

FIGURE 6.22. A. Soil map of the Arrapahoe Ridge, a storm beach. From Goodwin, 1987a, part of sheet 3. B. Cross-section of the Arrapahoe Ridge in the vicinity of A above. Generalized from Daniels et al., 1977.

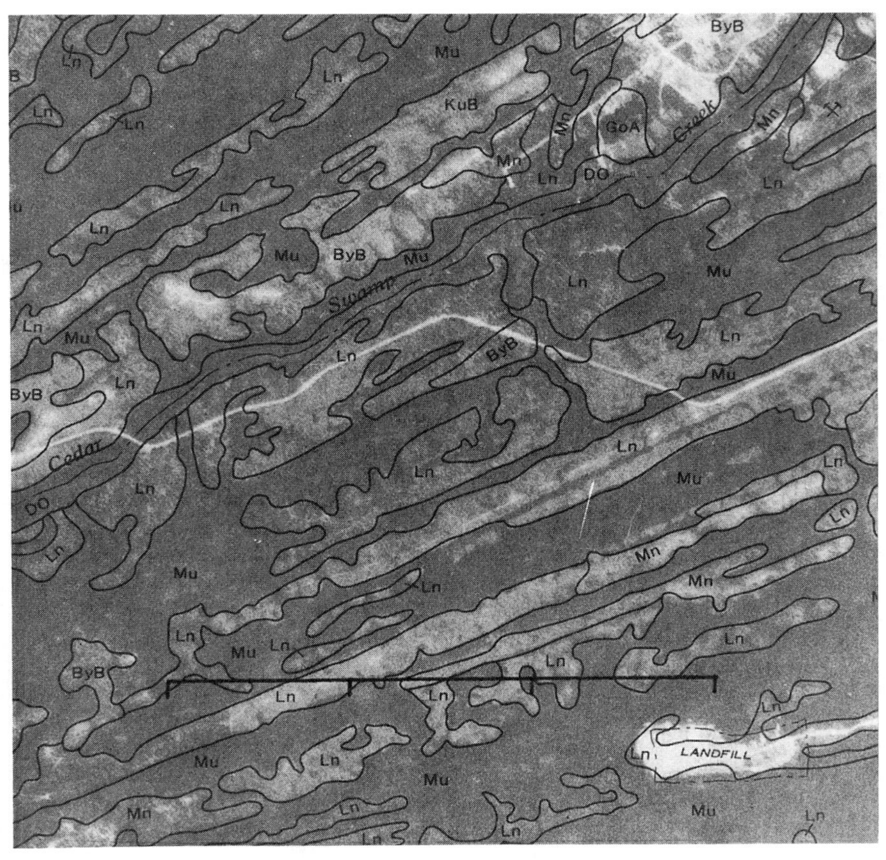

FIGURE 6.23. Soil map of prograding beach ridges south of the Neuse River, Craven County, North Carolina. From Goodwin, 1987b, part of sheet 11.

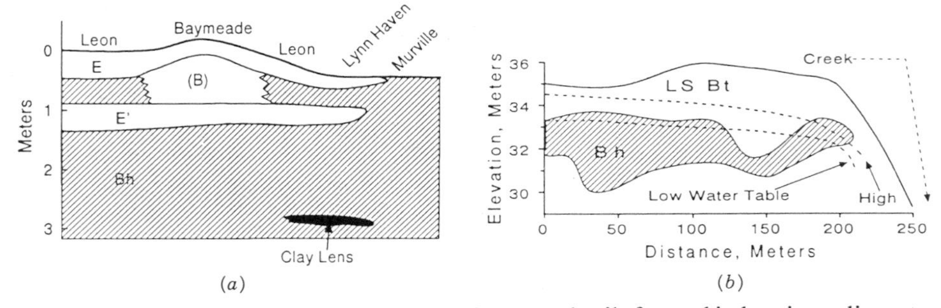

FIGURE 6.24. Relationship of spodic horizons and soils formed in barrier sediments to topography. A. Soils and horizonation. B. Spodic horizon thickness and distribution in relation to topography and location. From Daniels et al., 1976.

the beach face to water depths of 100 to 250 m. Inner-shelf sediments reflect the climate of their source area and the morphology of the immediate coast (Davis, 1983; Hayes, 1967). Polar shelves have gravelly, chloritic deposits, and temperate and arid tropical shelves have a sandy zone. Shelves abutting hot humid landmasses are muddy (Galloway & Hobday, 1983).

Estuaries and lagoons trap the coarser sediments from rivers before the finer material enters the shelf system. Currents transport only part of the finer sediments to the shelf through these traps. Much of the sand and gravel accumulating along present coasts is from older coastal sediments (Davis, 1983).

Beach and Shoreface

The beach and shoreface are the landward portion of the shelf system that extends from the beach seaward to wave base at 10 meters. The beach and shoreface environments have high but fluctuating energy. Storms transport shoreface sediments between the surf zone and the inner shelf even during moderate storms (Niedorda et al., 1985). Although these zones are narrow, the sediments are widespread in the geologic column.

The shore-zone sands are typically quartzose with zircon, tourmaline and rutile the major heavy minerals. The high-energy shore-zone environment winnows or removes most nonquartz minerals (Davis, 1983). Abrupt changes in mineralogy may occur where these sands interfinger with fluvial sediments (Galloway & Hobday, 1983).

The beach face environment is the swash zone. Heavy minerals concentrate in discrete laminae and often alternate with quartzose layers. Inverse grading is common with fine grains and heavy minerals changing upward into coarser sand. A berm, or subtle break in slope, separates the backshore from the foreshore. The backbeach is flat or slopes landward. Storm waves top the berm crest and deposit a thin sheet of sediment. Later wind-blown grains stick to the damp surface, initiating incipient dunes. Pebbles in the beach face environment are dominantly discoidal.

The shoreface is morphologically and dynamically distinct from the surf zone and the continental shelf (Niedorda et al., 1985). Galloway and Hobday (1983) subdivide the shoreface into the upper, middle and lower parts, each with characteristic textural properties. On most coasts, the upper shoreface deposits differ from those of the lower shoreface (Niedorda et al., 1985).

Upper shoreface environments are the landward part of the sequence with strong onshore, offshore and longshore currents. Coarse gravelly sediment concentrates toward the landward shoreface margin and has equant (equidimensional) pebbles. The coarsest textured sediments occur in the upper shoreface, but fine sands can occur in low-energy environments (Galloway & Hobday, 1983). Transgressive coasts have a seaward-fining and thinning blanket of sand.

Middle shoreface environments have powerful waves plus longshore and rip currents. Storm deposits are thicker and more lenticular than lower shoreface equivalents. Individual units can be a meter thick and hold sporadic basal gravel, shell or mudclast lags grading upward into massive sands.

The lower shoreface deposits are sand, silt or mud layers. The layers are a few centimeters thick and alternate irregularly or cyclically. These sediment properties develop from abrupt deposition followed by a period of bottom-dweller reworking. Each bed unit represents a storm event.

SHELVES

Several processes operate on the shelves, but tides, storm waves and current are most effective in moving sands and gravel. Other factors are temperature and salinity currents, seasonal convective sinking and overturning, and upwelling and downwelling of bottom waters (Galloway & Hobday, 1983).

Tidal and storm processes are most effective on the inner shelf. Density stratification and oceanic circulation are dominant processes on the outer shelf (Galloway & Hobday, 1983, Fig. 7-1, p. 144). Storm waves affect the outer-shelf to depths of more than 200 m, but fair weather waves have little effect. Shallow-water waves over a shoaling bottom cause a net onshore sediment transport. Storm waves erode beach sediments and deposit them on the shelf and shoreface. Density contrasts between salt and fresh water allow the fresh water plume from rivers to transport clays across the shelf.

Facies

Sediment supply and dominant processes control the texture and geometry of shelf facies. Tides and storms affect shelf sands, but muddy shelf deposits represent slow settling of suspended sediments and reworking by organisms. The storm- and tide-dominated sands are the coarse-grained members of the shelf deposits (Walker, 1979). The sand-ribbon facies are thin longitudinal bodies of sand (Kenyon, 1970) that have uniform width and spacing. The sand-ribbon facies probably are important in coarse basal transgressive layers, but volumetrically are not large (Galloway & Hobday, 1983). The sand-wave facies are transverse to the main tidal currents and form below wave action at a depth of about 18 m along the Dutch coast (McCave, 1971a).

Storm-induced flows also influence sand waves (Swift et al., 1973, 1977). Sand waves do not develop below tidal-current effectiveness, and those in deeper water are finer than shallow water equivalents (Galloway & Hobday, 1983). Sand ridges develop parallel to the major current flow and may be 40 meters high (Houbolt, 1968).

Finer shelf sediments surround some far offshore sand ridges. The shelf storm facies consist of shore-zone sediments deposited on the inner shelf during storms. Hayes (1967) described a graded gravelly sand bed developed during hurricane Carla that extends more than 24 km offshore. The storm deposit has a basal lag of shells gradationally overlain by sand. Burrowing destroyed the primary structures in less than 20 years (Morton, 1981). Storm deposits grade into shelf muds (Galloway & Hobday, 1983).

The mud facies on the mid- and outer-shelves may have secondary sand and shells. Bioturbated mud facies occur where depositional rates are high or where the bottom waters have little oxygen. Rivers and deltas of tropical areas supply abundant fines for transportation to the mid- or out-shelves (Galloway & Hobday, 1983; Hayes, 1967). Muds accumulate even under strong wave or current conditions if the suspended sediment concentrations are => 100 mg/l (McCave, 1971b). Biological activity, by extracting solids and ejecting them as faecal pellets, can accelerate the settling of clay particles by a factor of 10 (Drake, 1976; Johnson, 1975). Waves and tidal currents inhibit deposition at suspended sediment concentrations of <100 mg/l. The sediment concentrations on most shelves are 1 mg/l or less, so mud usually accumulates only in the deeper or protected areas where waves are not effective.

Transgressive-Regressive Facies The depositional phase determines the succession of shelf sediments (Curray, 1969; Galloway & Hobday, 1983; Winkler, 1980). Transgression reduces the contribution of sediment from rivers because the estuaries and tidal flats are efficient sediment traps. Transgressive systems usually exhibit a fining-upward sequence. The basal-gravel lag from wave erosion passes upward through sheet-like transgressive sands to finer sediments from successively deeper water sediments (Galloway & Hobday, 1983). Transgressive systems can have thick sediments along tectonically active continental margins, but on stable shelves rarely exceed 300 feet.

Transgressive tide-dominated sediments have basal gravel, massive sands and cross-bedded sands overlain by finer sediments. Sand ridges and tidal sand waves may be recognizable in the basal sands, but currents and burrowing organisms rework the upper surfaces. Fossiliferous bioturbated silts and muds from the outer-shelf overlie the sands (Fig. 6.25; Galloway & Hobday, 1983).

Transgressive storm-dominated shelf sediments also fine upward and are similar to the tide-dominated shelf deposits. The major difference is within the thin, gravelly, bioturbated shore-zone deposit overlying the subjacent unit (Galloway & Hobday, 1983).

A climax shelf deposit is a seaward-thickening prism of sediment. It grades from clean nearshore sands, through burrowed clayey inter-shelf sands, to mixtures of clay, silt and sand, and eventually to outer-shelf muds (Swift, 1969). Thus, prograding shelf sediments usually coarsen upward from shelf muds to cross-bedded or storm-graded sands.

Storm-dominated prograding shelf facies are the approximate reverse of the transgressive succession (Brenner, 1980; Moores, 1976). Inner-shelf sands occur, but much of the total sediment is a bioturbated, shelly mud. Shelves with anoxic bottom waters have laminated carbonaceous muds. The upward

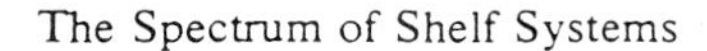

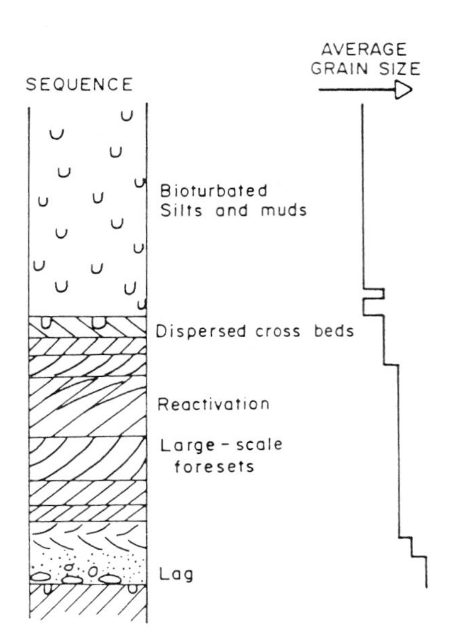

FIGURE 6.25. Vertical sequence of a transgressive tide-dominated shelf. From Galloway and Hobday, 1983, Fig. 7-14, p. 159. Reprinted by permission from Springer-Verlag and W.E. Galloway.

coarsening sequence changes from the basal storm sands into shore-zone sands (Fig. 6.26; Brenner, 1980; Moores, 1976).

Mixed-energy shelf sediments reflect a combination of storms, tides and possibly ocean currents. Several upward-coarsening sand units that parallel the shoreline overlie thick basal shelf muds (Brenner, 1980; Moores, 1976). Aggradational shelves can be exceedingly complex, having facies from the outer-shelf to tidal flat and estuarine deposits.

SUMMARY

Sediments from transitional depositional systems are widespread soil materials, but preservation of the entire depositional surface is unlikely. Delta organic and bay sediments may maintain their original shape, but the nearshore facies and the prodelta front may not. Preservation of estuarine or back-barrier flats and barrier-beach ridges is common. Considerable modification of beach and shoreface facies is the rule.

Exposure to soil-forming processes of the transitional systems occurs only in erosional landscapes. Surface form may be of little value in predicting material texture under these conditions. In sediments of major extent and thickness, sequences of materials have textures related to their original positions in the delta or barrier systems. For example, a vertical section through a prograding delta system will coarsen upward. The basal prodelta muds change upward through nearshore sands to distributary outlet gravel.

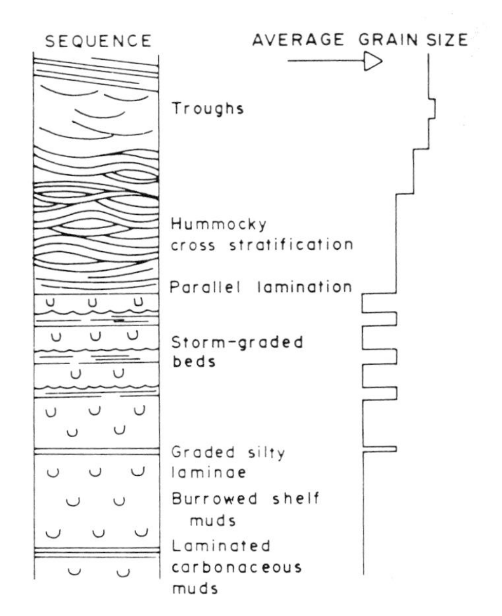

FIGURE 6.26. Vertical sequence of a prograding, storm-dominated shelf. From Galloway and Hobday, 1983, Fig. 7-17, p. 162. Reprinted by permission from Springer-Verlag and W.E. Galloway.

Estuarine systems will be coarsest at the shoreline and become finer toward the center. Coarser materials near stream channels can interrupt this sequence. Dissected barrier systems will have a linearity to the distribution of any one facies. The sediments may coarsen or fine upward, depending on the regressive or transgressive nature of the sediments. Inlet deposits, with their complex and abrupt changes, may dominate. The beach and shoreface sediments have characteristic textures within a system.

Soil scientists may not recognize each environment, so geologic maps that emphasize facies will help one predict the probable textures and areal extent of the unit. The most important thing to remember about any depositional system is that it has order. The system is not random. Therefore, one can predict what will be found and how it relates to associated materials. Useful predictions are not always easy to make in every landscape, but they are necessary if field work is to have application beyond the sample site.

ADDITIONAL READING

Elliott, T. (1986). Deltas. In *Sedimentary Environments and Facies* (pp. 113–154). Ed. by H.G. Reading. Boston: Blackwell.

Johnson, H.D. and C.T. Baldwin. (1986). Shallow siliciclastic seas. In *Sedimentary Environments and Facies* (pp. 229–282). Ed. by H.G. Reading. Boston: Blackwell.

Elliott, T. (1986). Siliciclastic shorelines. In *Sedimentary Environments and Facies* (pp. 115–188). Ed. by H.G. Reading. Boston: Blackwell.

REFERENCES

Barwis, J.H and M.O. Hayes. (1979). Regional patterns of modern barrier island and tidal inlet deposits as applied to paleoenvironmental studies. In *Carboniferous Depositional Environments in the Appalachian Region* (pp. 472–498). Ed. by J.C. Ferm and J.C. Horne. Univ. South Carolina, Carolina Coal Group.

Bates, C.C. (1953). *Amer. Assoc. Petrol. Geol. Bull.*, 37:2119–2162.

Boothroyd, J.C. (1985). Tidal inlets and tidal deltas. In *Coastal Sedimentary Sequences* (pp. 445–532). Ed. by E.A. Davis. New York: Springer-Verlag.

Brenner, R.L. (1980). *Amer. Assoc. Petrol. Geol. Bull.*, 64:1223–1244.

Coleman, J.M. and S.M. Gagliano. (1964). *Gulf Coast Assoc. Geol. Socs. Trans.*, 14:67–80.

Coleman, J.M. and L.D. Wright. (1975). Modern river deltas: Variability of processes and sand bodies. In *Deltas, Model for Exploration* (pp. 99–149). Ed. by M.L. Broussard. Houston, TX: Houston Geol. Soc.

Cronin, L.E. (Ed.) (1975). *Estuarine Research, Vol. 2.* New York: Academic Press.

Curray, J.R. (1969). Shore zone sand bodies: Barriers, cheniers and beach ridges. In *The New Concepts of Continental Margin Sedimentation* (pp. JC-11-1 to JC-11-18). Ed. by D.J. Stanley. Am. Geol. Inst.

Daniels, R.B., E.E. Gamble, and C.S. Holzhey. (1975). *Soil Sci. Soc. Amer. Proc.*, 39:1177–1181.

Daniels, R.B., E.E. Gamble, and C.S. Holzhey. (1976). *SE Geo.*, 13:61–75.

Daniels, R.B., E.E. Gamble, and W.H. Wheeler. (1977). *SE Geol.*, 18:231–247.

Davis, R.A. (1983). *Depositional Systems*. Englewood Cliffs, NJ: Prentice-Hall.

Drake, D.E. (1976). Suspended sediment transport and mud deposition on continental shelves. In *Marine Sediment Transport and Environmental Management* (pp. 127–158). Ed. by D.J. Stanley and D.J.P. Swift. New York: Wiley.

DuBar, J.R. (1971). *Neogene Stratigraphy of the Lower Coastal Plain of the Carolinas*. 12th Annual Field Conference, Field guidebook, mimeographed. Myrtle Beach, SC. Atlantic Coastal Plain Geological Association.

Edwards, J.M. and Frey, R.W. (1977). *Senckenberg. Marit.*, 9:215–259.

Elliott, T. (1975). Deltas. In *Sedimentary Environments and Facies* (pp. 97–142). Ed. by H.G. Reading. New York: Elsevier.

Fisher, W.L., L.F. Brown, A.J. Scott, and J.H. McGowen. (1969). *Delta Systems in the Exploration for Oil and Gas—A Research Colloquium*. Austin, TX: Bur. Econ. Geol., Univ. of Texas.

Frazier, D.E. (1967). *Gulf Coast Assoc. Geol. Soc. Trans.*, 17:287–315.

Frey, R.W. and P.B. Basan. (1985). Coastal salt marshes. In *Coastal Sedimentary Environments* (pp. 225–289). Ed. by R.A. Davis, Jr. New York: Springer-Verlag.

Friedman, G.M. and J.E. Sanders. (1978). *Principles of Sedimentology*. New York: Wiley.

Galloway, W.E. (1975). Process framework for describing the morphologic and stratigraphic evolution of deltaic depositional systems. In *Deltas* (pp. 87–98). Ed. by M.L. Broussard. Houston, TX: Houston Geol. Soc.

Galloway, W.E. and D.K. Hobday. (1983). *Terrigenous Clastic Depositional Systems*. New York: Springer-Verlag.

Glaser, J.D. (1978). *J. Geol.*, 86:283–297.

Goldsmith, V. (1985). Coastal dunes. In *Coastal Sedimentary Environments* (pp. 303–370). Ed. by R.A. Davis, Jr. New York: Springer-Verlag.

Goodwin, R.A. (1987a). *Soil Survey of Pamlico County, North Carolina.* USDA Soil Conservation Service.

Goodwin, R.A. (1987b). *Soil Survey of Carteret County, North Carolina.* USDA Soil Conservation Service.

Hayes, M.O. (1967). *Mar. Geol.*, 5:111–132.

Hayes, M.O. (1975). Morphology of sand accumulation in estuaries. In *Terrigenous Clastic Depositional Environments: Am. Assoc. Petrol. Geol. Field Course* (pp. 32–111). Ed. by M.O. Hayes and T.W. Kana. Univ. South Carolina, Tech. Rept., No. 11CRD, Pt. 1.

Holzhey, C.S., R.B. Daniels, and E.E. Gamble. (1975). *Soil Sci. Soc. Amer. Proc.*, 39:1182–1187.

Houbolt, J.J.H.C. (1968). *Geol. Mijn.*, 47:245–273.

Hoyt, J.M. and J.V. Henry. (1967). *Geol. Soc. Amer. Bull.*, 78:77–86.

Johnson, H.D. (1975). Shallow siliciclastic seas. In *Sedimentary Environments and Facies* (pp. 207–258). Ed. by H.G. Reading. New York: Elsevier.

Kenyon, N.H. (1970). *Mar. Geol.*, 9:25–39.

Kraft, J.C. (1971). *Geol. Soc. Amer. Bull.*, 82:2131–2158.

Kraft, J.C. and C.J. John. (1979). *Amer. Assoc. Petrol. Geol. Bull.*, 63:2145–2163.

Larsonneur, C. (19XX). Tidal deposits, Mont Saint-Michel Bay, France. In *Tidal Deposits* (pp. 21–30). Ed. by R.N. Ginsburg. New York: Springer-Verlag.

Lauff, G.H. (Ed.) (1967). *Estuaries.* Amer. Assoc. Adv. Sci. Publ. 83.

Markewich, H.W. et al. (1986). *Soil Development and its Relation to the Ages of Morphostratigraphic Units in Horry County, South Carolina.* U.S. Geol. Survey Bull. 1589.

McCave, I.N. (1971a). *Mar. Geol.*, 10:199–225.

McCave, I.N. (1971b). *J. Sediment. Petrol.*, 41:89–96.

Meckel, L.D. (1975). Holocene sand bodies in the Colorado Delta Area, Northern Gulf of California. In *Deltas* (pp. 87–98). Ed. by M.L. Broussard. Houston, TX: Houston Geol. Soc.

Moores, C.N.K. (1976). Introduction to physical oceanography and fluid dynamics of continental margins. In *Marine Sediment Transport and Environmental Management* (pp. 7–21). Ed. by D.J. Stanley and D.J.P. Swift. New York: Wiley.

Morton, R.A. (1981). Formation of storm deposits by wind-forced currents in the Gulf of Mexico and North Sea (pp. 385–396). In *Int. Assoc. Sediment. Spec.*, Pub. 5.

Morton, R.A. and J.H. McGowen. (1980). *Modern Depositional Environments of the Texas Coast.* [Guidebook 20] Austin, TX: Bur. Econ. Geol., Univ. of Texas.

Moslow, T.F. and S.D. Heron. (1989). *Outer Banks Depositional Systems, North Carolina.* [Field Trip Guidebook T171] 28th International Geological Congress, American Geophysical Union, Washington, D.C.

Nichols, M.M. and R.B. Biggs. (1985). Estuaries. In *Coastal Sedimentary Environments* (pp. 77–173). Ed. by R.A. Davis, Jr. New York: Springer-Verlag.

Niedorda, A.W., D.J.P. Swift, and T.S. Hopkins. (1985). The shoreface. In *Sedimentary Environments* (pp. 533–615). Ed. by R.A. Davis, Jr. New York: Springer-Verlag.

Nummedal, D. et al. (1977). Tidal inlet variability—Cape Hatteras to Canaveral. In Proc. on *Coastal Sediments* 77, Amer. Soc. Civil Engrs. Charleston, SC, pp. 543–562.

Reinson, G.E. (1980). Barrier island and associated strandplain systems. In *Facies Models* (pp. 57–74). Ed. by R.G. Walker. *Geoscience Canada*, Reprint Series No. 1.

Roy, P.S., B.G. Thom, and L.D. Wright. (1980). *Sed. Geol.*, 26:1–19.

Swift, D.J.P. (1969). Evolution of the shelf surface, and relevance of modern shelf studies to the rock record. In *The New Concepts of Continental Margin Sedimentation: Application to the Geological Record* (pp. DS-7-1–DS-7-19). Ed. by D.J. Stanley. Washington, DC: Amer. Geol. Inst.

Swift, D.J.P. (1975). *Sedim. Geol.*, 14:1–43.

Swift, D.J.P., D.B. Duane, and R.F. McKinney. (1973). *Mar. Geol.*, 15:227–247.

Swift, D.J.P., TR.A. Nelson, J.F. McHone, B.W. Holiday, H. Palmer, and G.L. Shideler. (1977). *J. Sediment. Petrol.*, 49:1454–1474.

Thom, B.G., H.A. Poloxch, and G.M. Bowman. (1978). *Holocene Age Structure of Coastal Sand Barriers in New South Wales, Australia*. Dunstroon: Univ. New South Sales, Fac. Military Stud., Dept. Geogr. Royal Military College.

Walker, R.G. (1979). Shallow marine sands. In *Facies Models* (pp. 75–89). Ed. by R.G. Walker. *Geoscience Canada*.

Welder, F.A. (1959). *Processes of Deltaic Sedimentation in the Lower Mississippi River*. Coastal Stud. Inst., Tech. Rept., No. 12, Louisiana State Univ.

Winkler, C.D. (1980). Depositional phases in Late Pleistocene cyclic sedimentation. In *Middle Eocene Coastal Plain and Nearshore Deposits of East Texas: A Field Guide to the Queen City Formation and Related Papers* (pp. 46–66). Ed. by B.F. Perkins and D.K. Hobday. Gulf Coast Section, Soc. Econ. Paleont. Mineral.

Wright, L.D. (1985). River deltas. In *Coastal Sedimentary Environments* (pp. 1–76). Ed. by R.A. Davis. New York: Springer-Verlag.

Wright, L.D. and J.M. Coleman. (1974). *J. Geol.*, 82:751–778.

7 Volcanic Materials

INTRODUCTION

Volcanic terrains have a greater variety of rock-types than any other surface environment on earth (Cas & Wright, 1987). These terrains include lavas, explosive pyroclastic deposits, and deposits from a wide range of sedimentary processes that occur in volcanic terrains. Although several soil orders occur on volcanic materials, Andisols occur almost exclusively on pyroclastic volcanic deposits. Soils developed in pyroclastic and other fragmented volcanic materials occupy only about 0.8% of the earth's surface.

The central concept of an Andisol (ICOMAND, 1988) is that of a soil developing in volcanic ejecta (such as volcanic ash, pumice, cinders, lava) and/or in volcaniclastic materials, with short-range-order minerals (amorphous) or Al-humus complexes dominating the colloidal fraction. Under some environmental conditions, weathering of primary alumino-silicates in parent materials of non-volcanic origin may also lead to the formation of short-range-order minerals; some of these soils are also Andisols. The Soil Survey Staff (1990) defined the many properties of Andisols.

The emphasis in this chapter will be on pyroclastics and some of the processes responsible for their deposition. Figure 7.1 shows the distribution of a succession of pyroclastic-flow deposits from the May 18, 1980, and later eruptions of Mount St. Helens, Washington (Rowley et al., 1981). Volcanic terrains can have a complex soil pattern because different depositional units may have a wide textural range.

Few soil orders, except organic soils, have such a specific range of parent materials and depositional environments. Lavas and volcanic ejecta have a wide range of environments and mineralogy, even from the same vent. For one studying soils in these terrains it is important to understand the depositional processes of these materials and their areal variability.

KINDS OF VOLCANIC ROCKS

Classification of volcanic rocks is by their chemical or mineralogical properties. A simple classification uses silica for the initial division into silicic, intermediate, basic and ultrabasic classes (Cas & Wright, 1987). Table 7.1 is a classification based on chemical composition, and Figure 7.2 is one that uses mineralogy. For a more detailed mineralogy classification see Streckeisen

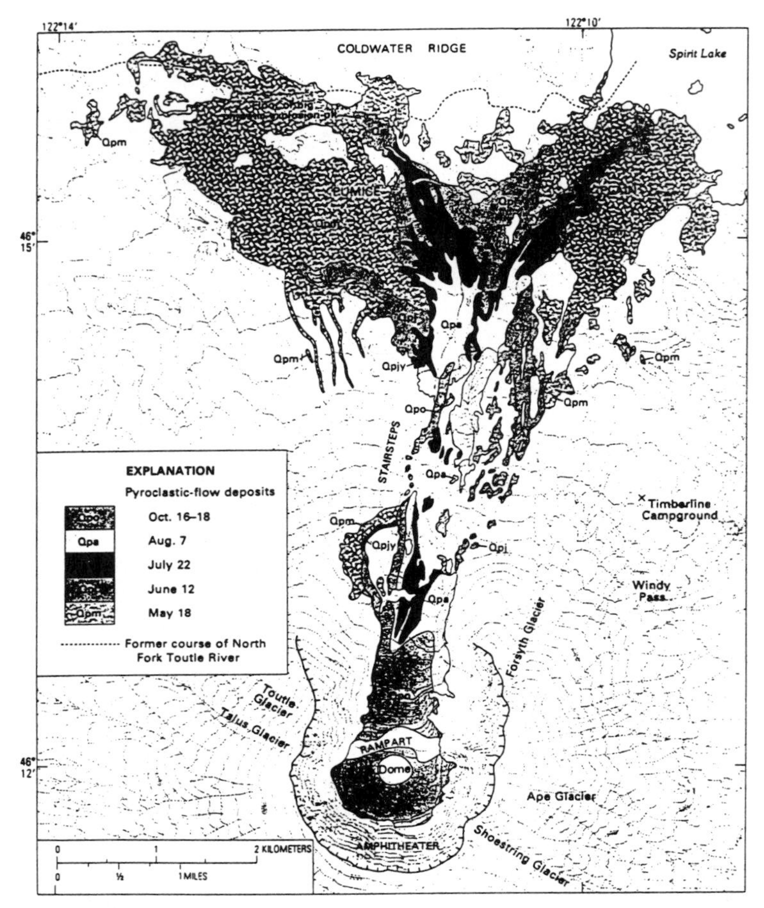

FIGURE 7.1. *Volcaniclastic deposits, excluding fallout ash, from the 1980 eruptions of Mount St. Helens.* From Rowley et al., 1981, Fig. 295, p. 503. Reprinted by permission from P.D. Rowley.

(1979) or Cas and Wright (1987). The processes that give fragmented volcanic deposits are discussed briefly because they are important soil-forming materials.

Fragmented Volcanic Rocks

Primary processes (eruption) and secondary surface processes such as weathering, erosion and mass-wastage produce fragmented volcanic rocks (Cas & Wright, 1987). Both the primary and secondary processes develop breccias, sand- and mud-sized aggregates. Particles expelled through volcanic vents during eruptions become pyroclastic fragments (Fisher & Schmincke, 1984).

TABLE 7.1. A Simple Chemical Classification for the Common Volcanic Rock Types

		Al_2O_3 saturation classes			
	SiO_2 (wt%)	Peraluminous*	Metaluminous†	Subaluminous‡	Peralkaline§
acid	>68	rhyolite or obsidian →			pantellerite, comendite
	63–68	rhyodacite →			
		dacite →			
			latite	trachyte ↓	↑ phonolite
intermediate	57–63		andesite		
	52–57		mugearite		
			tholeiitic basalt		
			hawaiite		
Basic	45–52		alkali basalt		
			basanite		
ultrabasic	<45		nephelinite, leucitite		

* Molecular $Al_2O_3 > (CaO + Na_2O + K_2O)$.
† Molecular $Al_2O_3 < (CaO + Na_2O + K_2O)$ and $Al_2O_3 > (Na_2O + K_2O)$.
‡ Molecular $Al_2O_3 \sim (Na_2O + K_2O)$.
§ Molecular $Al_2O_3 < (Na_2O + K_2O)$.

Note: Basaltic rocks cover a wide compositional range and can be further subdivided. For more-comprehensive chemical classification schemes, see Yoder and Tilley (1962), Green and Ringwood (1967) and Irvine and Baragar (1971).

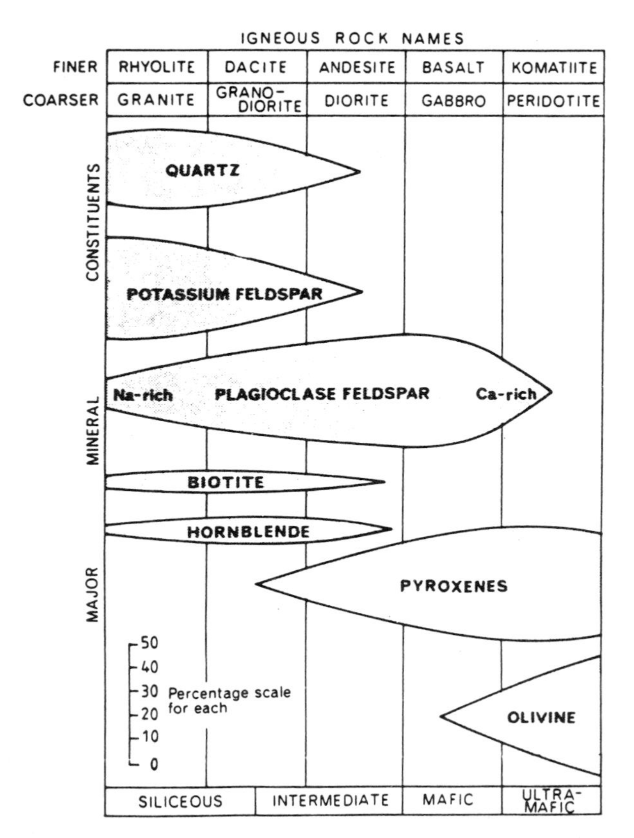

FIGURE 7.2. *Classification of intrusive and extrusive igneous rocks and their major components.* From Selby, 1985, Fig. 3.18, p. 55. Republished by permission from M.J. Selby.

Hydroclastic materials are pyroclasts from steam explosions at magma-water interfaces. Another mechanism is from rapid chilling and mechanical granulation of lava when it contacts water or water-saturated sediments (Fisher & Schmincke, 1984).

Terminology Volcanic fragments occur in other deposits and geologists have specific names for volcanic rocks fragmented by the various processes. *Epiclastic* rocks are from weathering and erosion of older volcanic rocks. *Autoclastic* rocks become fragmented during movement of lava or by gravity crumbling. *Alloclastic* rocks are from disruption of preexisting volcanic rocks by igneous processes beneath the earth's surface (Fisher & Schmincke, 1984). *Volcaniclastic* is a general term that includes all clastic volcanic materials. *Tephra* is a term for pyroclastic accumulations of any size (Fisher & Schmincke, 1984). Tables 7.2 and 7.3 give the classification of pyroclastic rocks or a mixture of pyroclastic and epiclastic rocks. Figure 7.3 gives the end member rock terms for pyroclastic rocks.

Scoria, cinders and pumice are common types that usually are lapilli (Fig. 7.3) or larger size. Pumice is white to brown, highly vesicular, silicic to mafic glass that usually will float on water. Scoria and cinders are essentially synonymous terms applied to opaque glass, usually of mafic composition, that readily sinks in water (Fisher & Schmincke, 1984). For a detailed discussion of properties of pyroclastic fragments and deposits see Chapter 5 of Fisher and Schmincke.

Pyroclastic Deposits Pyroclastic deposits are from fragmentation of magma and rock by explosion. They divide into three units based upon the mode of transport and deposition: falls, flows and surges (Fig. 7.4). Fall deposits develop by explosive ejection from a vent (Figs. 7.5 and 7.6). Column height and wind velocity and direction control the size, geometry and extent of the fall deposit (Cas & Wright, 1987). Fragment size ranges from bombs, >64 mm, to ash, <2 mm (Waitt et al., 1981). Fall deposits mantle the landscape (Fig. 7.3), are moderately well sorted, and thickness and particle size decrease downwind.

Pyroclastic Flow Deposits. The volume of pyroclastic flows ranges from 0.001 to 100 km^3, and the composition can change with flow size. Small to intermediate volume flows have rhyolitic to basaltic compositions and large-volume flows are rhyolitic to dacitic. There also is a close relationship between the kinds of fragments and the flow origin. Smaller flows produced by dome-collapse or explosions during dome formation may contain a large amount of poorly vesiculated products. Highly vesiculated materials are common in inter-

TABLE 7.2. Classification of Pyroclasts and Unimodal, Well-sorted Pyroclastic Deposits

Clast Size	Pyroclast	Pyroclastic Deposit: Mainly Unconsolidated: Tephra	Pyroclastic Deposit: Mainly Consolidated: Pyroclastic Rock
	Block, bomb	Agglomerate, bed of blocks or bomb, block tephra	Agglomerate, pyroclastic breccia
64mm			
	Lapillus	Layer, bed of lapilli or lapilli tephra	Lapilli tuff
2 mm			
	Coarse ash grain	Coarse ash	Coarse (ash) tuff
1/16 mm			
	Fine ash grain (dust grain)	Fine ash (dust)	Fine (ash) tuff (dust tuff)

After Schmid, 1981, Table 1, p. 42. Republished by permission from R. Schmid.

TABLE 7.3. Terms for Mixed Pyroclastic and Epiclastic Rocks

Pyroclastic		Tuffites Mixed Pyroclastic and Epiclastic	Epiclastic (Volcanic and/or Non-Volcanic)	Average Clast Size (mm)
Agglomerate, agglutinate pyroclastic breccia		Tuffaceous conglomerate tuffaceous breccia	Conglomerate, breccia	64
Lapilli tuff				
(Ash) tuff	coarse	Tuffaceous sandstone	Sandstone	2
	fine	Tuffaceous siltstone	Siltstone	1/16
		Tuffaceous mudstone, shale	Mudstone, shale	1/256
100%	75%		25%	0% by volume
(Increase)				
______	______	______	Pyroclasts	
				(Increase)
______	______	______	Volcanic + non-volcanic epiclasts-----------	

From Schmid, 1981, Table 2, p. 43. Republished by permission from R. Schmid.

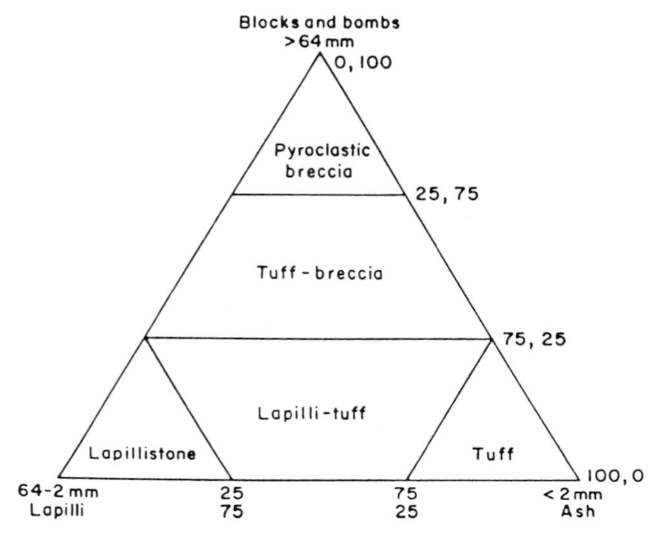

FIGURE 7.3. *Terms for mixtures and end member rocks of pyroclastic fragments.* From Fisher, 1966. Reprinted by permission from Elsevier Science Publications and R.V. Fisher.

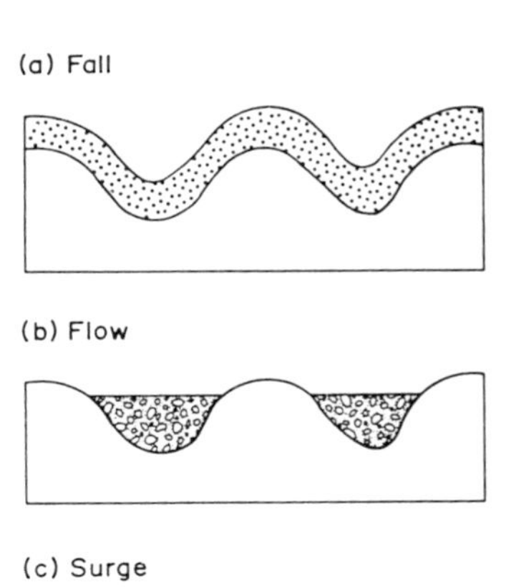

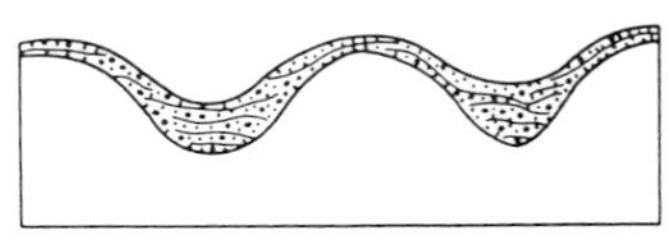

FIGURE 7.4. *Relationship of fall, flow and surge deposits to topography.* From Wright et al., 1980, Fig. 2, p. 317. Republished by permission from Elsevier Scientific Publishers and J.V. Wright.

mediate to large-volume flows (Fisher & Schmincke, 1984). Flow distance increases with flow volume (Smith, 1960).

Gravity-controlled surface flows of hot debris with high particle concentration regulate pyroclastic flow deposition. These deposits usually fill valleys and depressions (Fig. 7.3). Flow deposits thicken away from the source. They

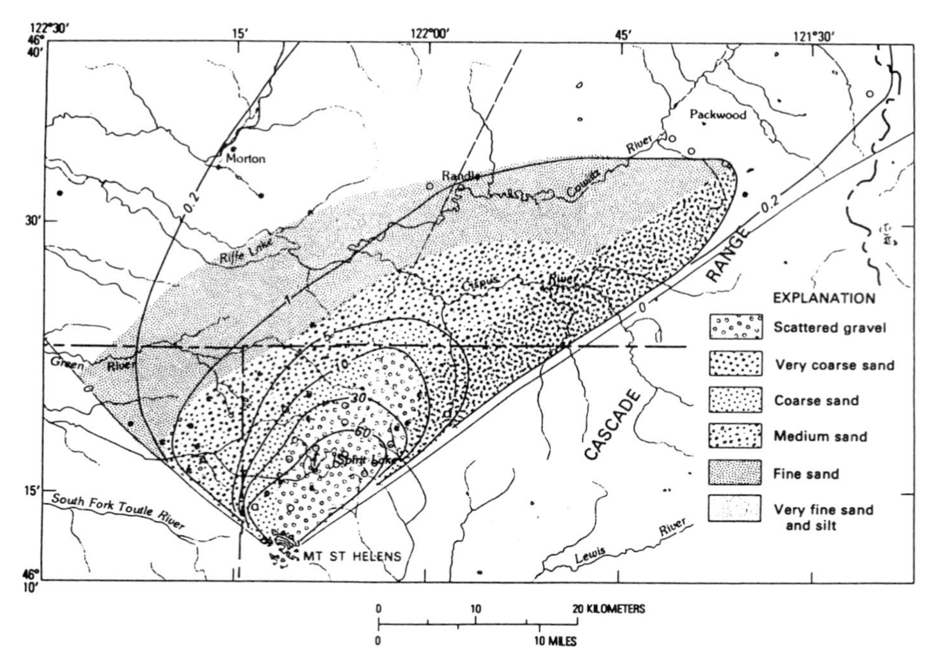

FIGURE 7.5. *Map of August 7 air-fall deposit showing median grains size.* Isopleths are of intermediate diameter of largest pumice fragments in millimeters. From Waitt et al., 1981, Fig. 373, p. 625. Republished by permission from R.B. Waitt.

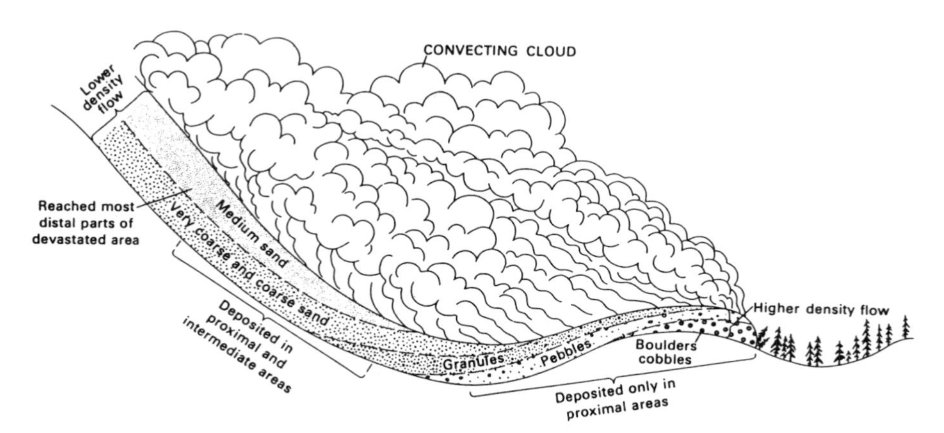

FIGURE 7.6. *Diagram of pyroclastic density flow* showing inferred maximum particle size and relative density. From Waitt, 1981, Fig. 270, p. 456. Republished by permission from R.B. Waitt.

may drain completely from upper slopes (Gorshkov, 1959). Small-volume flow may leave levees and larger rock fragments on both sides of a valley by draining down its center (Fisher & Schmincke, 1984). Beyond the mountain slopes, the flows spread out in fan-like lobes (Fig. 7.1). The flow deposits are massive and poorly sorted but may show grading of larger clasts (Cas & Wright, 1987). Deposition of these materials occurs at temperatures above 390 degrees C. The deposits have carbonized wood and a pink coloration from thermal oxidation of iron and zones of welded tuft. Figure 7.6 is a schematic of a pyroclastic flow. Figure 7.7 illustrates a composite stratigraphic column and the associated textural relations from a Mount St. Helens pyroclastic density flow and air-fall deposit.

Pyroclastic Surge Deposits. Surge deposits move along the surface as an expanded, turbulent, low-particle gas-solid dispersion (Cas & Wright, 1987). They mantle the topography, but also thicken in depressions. Well-sorted beds of surge material exist, but the range of sorting across beds can be large. Surge deposits usually have better sorting, are finer-grained and have more distinct bedding than flow deposits (Fisher & Schmincke, 1984). Surge deposits may predate a pyroclastic flow. They develop from eruption column collapse and from infolding air at the front of a pyroclastic flow. They form at the base of the eruption column and travel outward at initial velocities as high as 100 m s^{-1} (Moore, 1967).

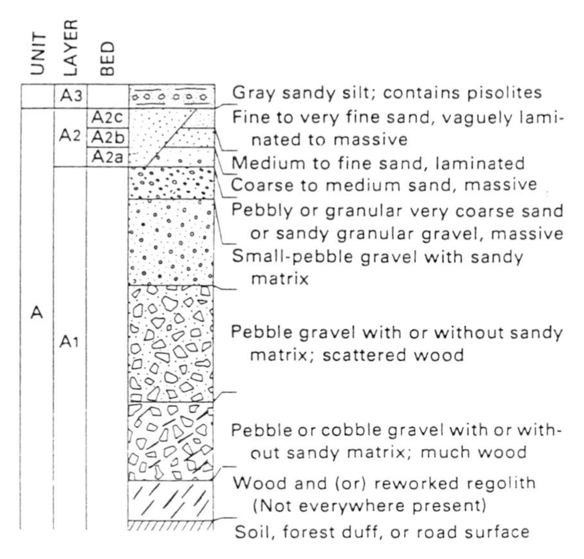

FIGURE 7.7. *Composite stratigraphic column of Mount St. Helens pyroclastic density flow and air-fall deposits* (Unit A). From Waitt, R.B. Jr, 1981, Fig. 258, p. 441. Republished by permission from R.B. Waitt.

Pyroclastic Fall Deposits. Fall deposits result from vertically directed eruptions and from the ash clouds associated with pyroclastic flows and surges (Fig. 7.5). Sorting of particles within the eruption column is by size and density, but wind does not effect the larger particles because they follow ballistic trajectories (Fisher & Schmincke, 1984). Turbulence temporarily suspends heavy particles in the eruption cloud, but atmospheric winds carry many lighter particles from the site. Thus, there is considerable zonation of the size of fall material away from the source. Although fragments with small fall velocities may circle the earth several times (Lamb, 1970), many of these particles remain near the source. Accretionary lapilli fall near the source because the very small particles react with moisture in the eruption cloud (Brazier et al., 1982). Several processes sort air-fall deposits. Rainfall, trapping in a heavy fall of coarse-grained ash and lapilli and electrostatic attraction of clusters, helps sort air-fall deposits (Fisher & Schmincke, 1984; Sorem, 1982). The maximum size and sorting usually decrease with distance from source, although poorly sorted deposits occur close to the source. Downwind changes in total composition of the ash layer result from the different but overlapping grain size of crystals, lithic or dense vitric fragments and pumice, including glass shards (Fisher & Schmincke, 1984).

Tephra sheets have circular patterns from low eruption columns during calm winds and fan-shaped patterns from high eruption columns and strong unidirectional winds. Contrasting winds at different altitudes can distribute contemporaneous sheets in opposite directions (Wilcox, 1959). Tephra sheets usually thin systematically from the source, but compaction, erosion and topography cause considerable variability (Fisher & Schmincke, 1984).

The above are gross generalizations of an extremely complex suite of soil-forming materials. We recommend the U.S. Geological Survey Professional Paper on the 1980 eruptions of Mount St. Helens as an excellent source of information on volcanic processes (Lipman & Mullineaux, 1981). The other major texts cited are also excellent, but none can give the detail of the professional paper by Lipman and Mullineaux. The professional paper also is an excellent source of additional references.

PYROCLASTIC MATERIALS AND SOILS

Thorpe and Smith (1949) recognized that soils in pyroclastic materials have unique properties and coined the name *Ando* (Wada, 1985). Andisols have high water-holding capacity, "smeary" (slippery but non-sticky) consistence, low bulk densities, high void ratios, and may collapse irreversibly when dried. Chemical properties include a high affinity for organic matter, pH-dependent charge, weak ionic absorption strength, and high phosphate-fixation capacities (Wada & Harward, 1974).

Secondary minerals in Andisols (allophane, immogolite, and halloysite) have distinctive properties. These minerals have tubular (halloysite) and thread-like (immologlite) structures rather than the platy structure of kaolinite, vermiculite, illite or montmorillonite. Wada (1989) gives a detail discussion of secondary minerals associated with Andisols.

Papers by Maeda et al. (1977), Wada (1985), and Wada and Harward (1974) discuss in detail the unique properties of soil formed in pyroclastic materials. For a detailed discussion of classification of Andisols see Buol et al. (1989, pp. 317–323). The above citations are excellent sources for the abundant literature on Andisols. Much of the literature describing the properties of Andisols is of Japanese origin.

Although an abundant literature exists on the properties of Andisols, few studies have detailed the relationships between the various kinds of pyroclastic materials and soils on a landscape. Few studies treat the unique geology and soil as one system.

REFERENCES

Brazier, S., A.N. Davis, H. Sigurdsson, and R.S.J. Sparks. (1982). *J. Volcanol. Geotherm. Res.*, 14:335–359.

Buol, S.W., F.D. Hole, and R.J. McCracken. (1989). *Soil Genesis and Classification*. 3rd Ed. Ames: Iowa State Univ. Press.

Cas, R.A.F. and J.V. Wright. (1987). *Volcanic Successions*. London: Allen & Unwin.

Fisher, R.V. (1966). *Earth-Sci. Rev.*, 1:287–298.

Fisher,R.V. and H.-U. Schmincke. (1984). *Pyroclastic Rocks*. Berlin: Springer-Verlag.

Gorshkov, G.S. (1959). *Bull. Volcanol.*, 20:77–109.

ICOMAND. [International Committee of the Classification of Andisols] (1988). *Circ.*, 10.

Lamb, H.H. (1970). *Phil. Trans. Roy. Soc. London*, A 266:425-533.

Lipman, P.W. and D.R. Mullineaux. (Eds.) (1981). *The 1980 eruptions of Mount St. Helens*. U.S. Geol. Survey Prof. Paper 1250.

Maeda, T., H. Takenada, and B.P. Warkentin. (1977). *Adv. Agron.*, 29:229–264.

Moore, J.G. (1967). *Bull. Volcanol.*, 30:337–363.

Rowley, P.D., M.A. Kuntz, and N.S. MaCleod. (1981). Pyroclastic flow deposits. In *The 1980 Eruptions of Mount St. Helens* (pp. 489–512). Ed. by P.W. Lipman and D.R. Mullineaux. U.S. Geol. Survey Prof. Paper 1250.

Schmid, R. (1981). *Geology*, 9:41–43.

Selby, M.J. (1985). *Earth's Changing Surface*. Oxford: Clarendon Press.

Smith, R.L. (1960). *Geol. Soc. Amer. Bull.*, 71:795–842.

Soil Survey Staff. (1990). *Keys to Soil Taxonomy*. [SMSS Tech. Mono. 19.] 4th ed. Blacksburg, VA: VPI.

Sorem, R.K. (1982). *J. Volcano. Geotherm. Res.*, 13:63–71.

Streckeisen, A. (1979). Classification and nomenclature of volcanic rocks, lamphorphores, carbonatites, and melilitic rocks: Recommendations and suggestions of the IUGS Subcommission on the Systematics of Igneous Rocks. *Geology*, 7:331–335.

Thorpe, J. and G.D. Smith. (1949). *Soil Sci.*, 67:117–126.

Wada, K. (1985). The distinctive properties of Andisols. In *Advanced Soil Science 2* (pp. 173–229). Ed. by B.A. Steward. New York: Springer-Verlag.

Wada, K. (1989). Allophane and Imogolite. In *Minerals in Soil Environments* (pp. 1052–1081). Ed. by J.B. Dixon and S.B. Weed. SSSA Book Series No. 1.

Wada, K. and M.E. Harward. (1974). *Adv. Agron.*, 26:211–260.

Waitt. R.B., Jr. (1981). Devastating pyroclastic density flow and attendant air fall of May 18—Stratigraphy and sedimentology of deposits. In *The 1980 Eruptions of Mount St. Helens* (pp. 439–458). Ed. by P.W. Lipman and D.R. Mullineaux. U.S. Geol. Survey Prof. Paper 1250.

Waitt, R.B., Jr., V.L. Hansen, A.M. Sarna-Wojcicki, and S.H. Wood. (1981). Proximal air-fall deposits of eruptions between May 24 and August 7, 1980—Stratigraphy and field sedimentology. In *The 1980 Eruptions of Mount St. Helens* (pp. 617–628). Ed. by P.W. Lipman and D.R. Mullineaux. U.S. Geol. Survey Prof. Paper 1250.

Wilcox, R.E. (1959). Some effects of recent volcanic ash falls with especial reference to Alaska. *U.S. Geol. Survey Bull.*, 1028-N:409–476.

Wright, J.V., A.L. Smith, and S. Self. (1980). *J. Volcanol. Geotherm. Res.*, 8:315–336.

8 Saprolite

Saprolite is the weathering product of the underlying bedrock that has decomposed in place (Pavich et al., 1989). It can be clayey to sandy, has not moved, and it retains the original rock structure (Becker, 1895). Saprolite is soft enough to be dug or chopped with a hand shovel. It is a widespread soil material in the Appalachian, Piedmont, Valley and Ridge Province, Cumberland Plateau, Sierra Nevada Mountains and other areas with deeply weathered bedrock (Hack, 1982; O'Brien & Buol, 1984; Ruxton & Berry, 1957).

Saprolite develops because groundwater reacts with minerals in the bedrock and weathered regolith (Bricker et al., 1968). Groundwater flow and weathering remove the easily soluble minerals with little change in original rock volume (Cleaves et al., 1974; Pavich, 1986; Pavich et al., 1989).

Soil scientists have studied the saprolite and soils developed from saprolite for many years. An extensive literature examines soil and saprolite relations in the southern Piedmont and similar areas (Cady, 1950; Calvert et al., 1980; Dadgari, 1982; Nutter & Otten, 1969; Otten, 1981; Rebertus & Buol, 1986; Rebertus et al., 1986; Rice et al., 1985; Simpson, 1986; Smith, 1986). A much more limited data base is available on saprolite thickness by landscape position (Froelich, 1975; Froelich & Heironimus, 1977; Pavich et al., 1989).

REGOLITH PROPERTIES

Geologists have extensively and intensively studied the properties of the weathered mantle in the southern Piedmont and have divided it into several zones (Cleaves & Costa, 1974; Pavich et al., 1989). The zonation from the bottom up is unweathered rock, weathered rock, saprolite, massive subsoil and soil (Fig. 8.1; Pavich et al., 1989). The unweathered rock lacks visible alteration of minerals, but iron stains the joint faces. Weathered rock has about 10% or more hard core stones that ring when struck with a hammer. Weathered feldspars and biotite occur in this zone. The overlying saprolite makes up most of the weathering profile in quartzofeldspathic rock (Pavich et al., 1989).

Saprolite is the weathering product of the underlying bedrock that has decomposed in place. To pedologists, this is the C horizon of the overlying soil. Clay content of the saprolite may increase slightly from its base to the overlying massive subsoil, or remain somewhat constant throughout (Pavich et al., 1989, Plate 3).

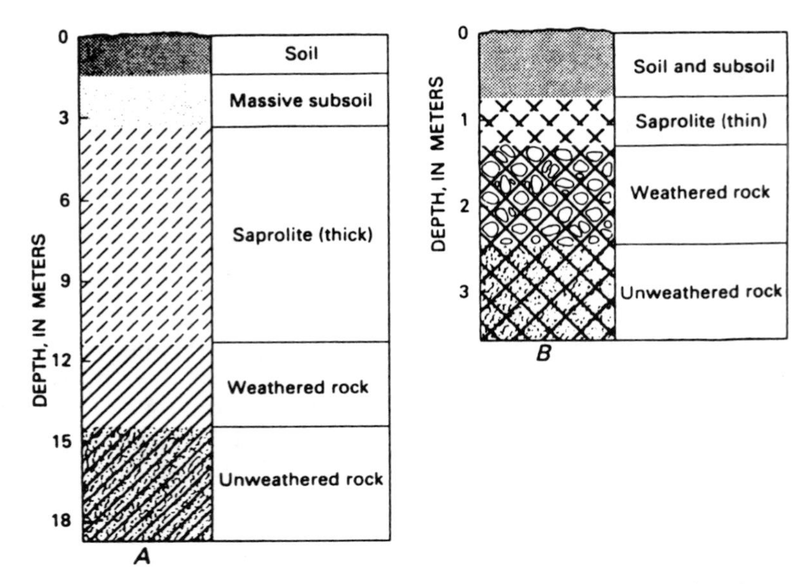

FIGURE 8.1. *Weathering profiles of crystalline rocks in Virginia.* A. Foliated metasedimentary and granitic rocks. B. Massive igneous rock such as diabase. From Pavich et al., 1989, Fig. 2, p. 5; originally modified from Langer, 1978. Republished by permission from M.J. Pavich.

The massive subsoil between the saprolite and the soil B horizon (Pavich et al., 1989) equals the BC and CB horizons of the pedologist. Some workers may call it C horizon and not differentiate it from the deeper saprolite with rock structure. The clay content increases slightly in the massive subsoil but is less than in the overlying B horizon. The soil-zone of Pavich et al. is the A, E and B horizons of the pedologist. Characterization of soils formed in saprolite should use soil science criteria. Pavich and associates' (1989) interests are in the entire weathered section, not just the layers we call soil.

AREAL DISTRIBUTION AND THICKNESS

Both topographic position and rock-type exert major controls on thickness and properties of the regolith. The thickest weathered zones on igneous and metamorphic rock occur beneath the broad gentle interfluve summits. The thinnest weathered zones are on the steeper side-slopes where erosion is active (Fig. 8.2). In the Occoquan granite, the saprolite is thickest beneath the divides and slopes gently toward the stream. The steep valley slopes truncate the saprolite and expose weathered and unweathered bedrock (Pavich, 1986, Fig. 4). A 1 m thick residual soil caps all but the steeper valley slopes. Soils on the steep valley slopes are in colluvium and residuum.

The thickest weathered zones occur on schistose, gneissic and granitic rocks and the thinnest zones on ultramafic rocks (Figs. 8.1 and 8.3; Cleaves et al., 1970; Froelich, 1975). Although difficult to quantify, the major kind of bedrock can be predicted from the topography alone by many experienced soil mappers. It is logical that the thick weathering zones underlying granitic landscapes and the thin regolith of the more mafic and ultramafic rock would cause distinctive topography.

Characteristics

Saprolite has a vertical chemical and mineralogical zonation. Weathering removed Ca and Na from the upper 10 m of the saprolite, but these elements increased in the underlying weathered and unweathered rock (Fig. 8.4A). Most of the chemical loss from the original rock is Ca, Na, and Si, which correlates with petrographic data on alteration and loss of plagioclase feldspar (Fig. 8.4B). Biotite and pyrite alter in the weathered rock-zone, but the mass loss is small (Pavich, 1986). Density decreases from the bedrock upward and reaches a minimum in the upper part of the saprolite before increasing again in the lower solum (Fig. 8.5). O'Brien and Buol (1984) show similar changes.

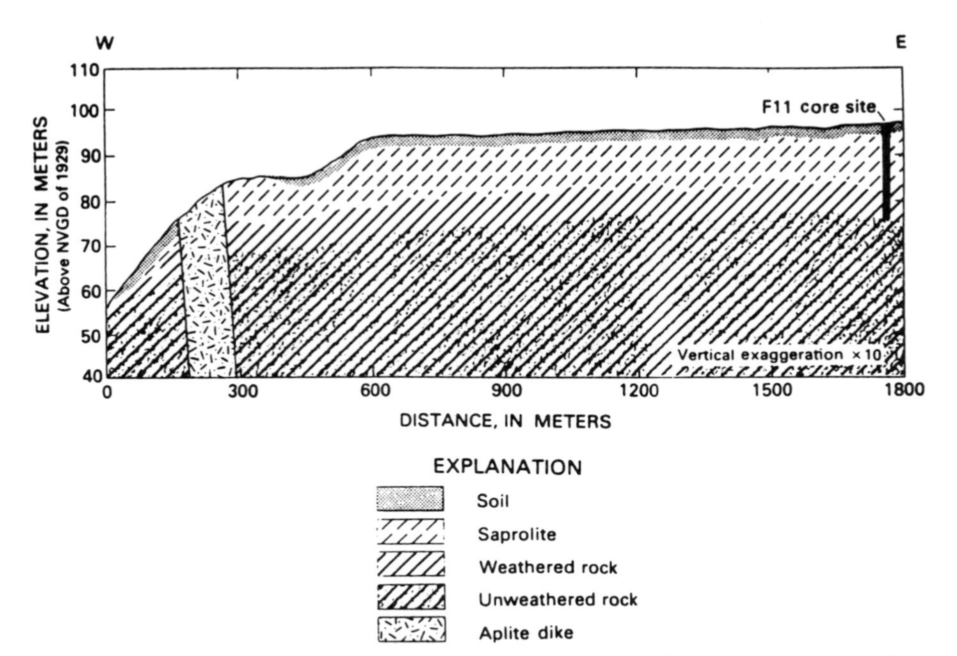

FIGURE 8.2. *Areal distribution of weathered zones in a granite landscape.* From Pavich et al., 1989, Fig. 10, p. 20. Republished by permission from M.J. Pavich.

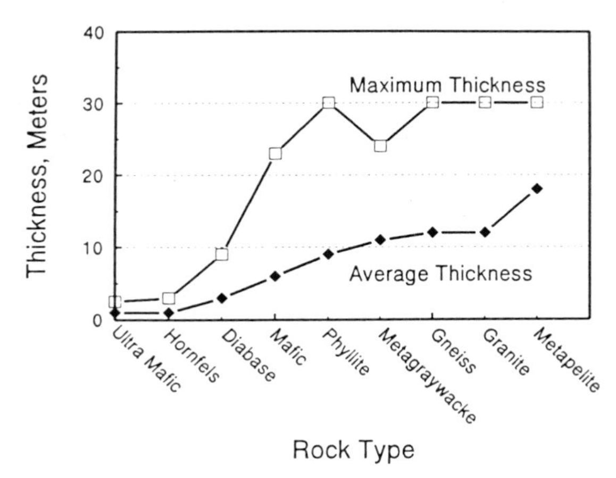

FIGURE 8.3. *Relationship between average and maximum thickness of regolith and rock-type.* Redrawn from Pavich et al., 1989. Plate 1. Ultramafic = serpentinite, gabbro. Hornfels = metamorphosed shale, siltstone, sandstone. Diabase = intrusive Triassic dikes. Mafic = metaigneous, metavolcanic, and volcaniclastic greenstones. Phyllite = phyllite, metasiltstone and phyllitic slate. Metagraywacke = impure quartzite, metagraywacke, schist. Gneiss = schistose gneiss, schist. Granite = granite, adamellite, granodiorite. Metapelite = pelitic schist, mica schist, metagraywacke. Republished by permission from M.J. Pavich.

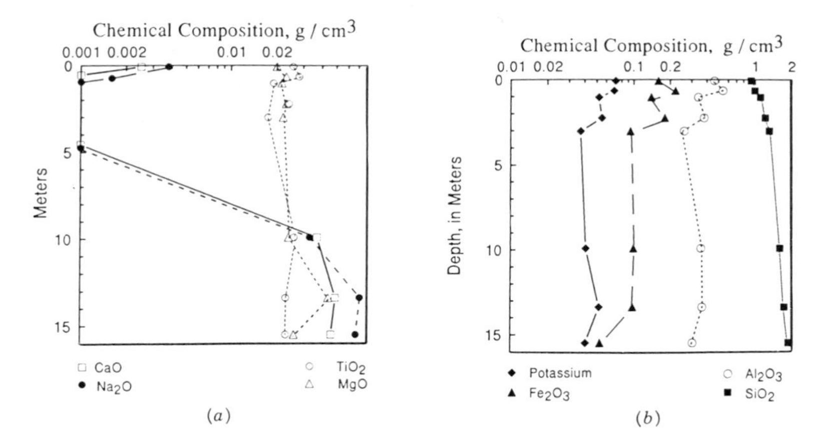

FIGURE 8.4. *Vertical distribution of oxides in a metapelite weathered section*. A. Calcium, Sodium, Titanium & Magnesium; B. Potassium, Iron, Aluminum and Silica. From Pavich et al., 1989, Plate 3.

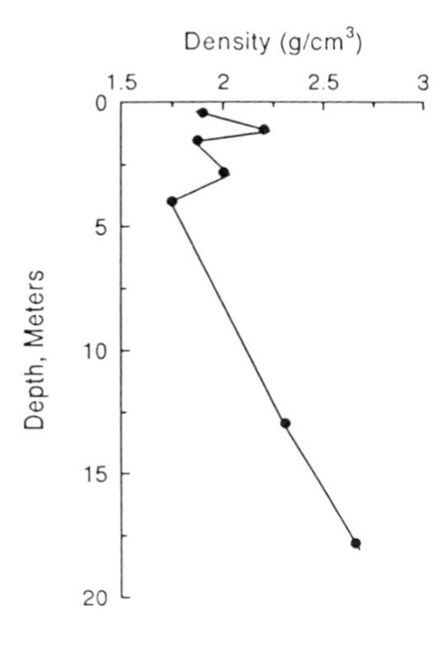

FIGURE 8.5. *Vertical changes in density from unweathered rock to soil in a metapelite weathered section.* From Pavich et al., 1989, Plate 3. Republished by permission from M.J. Pavich.

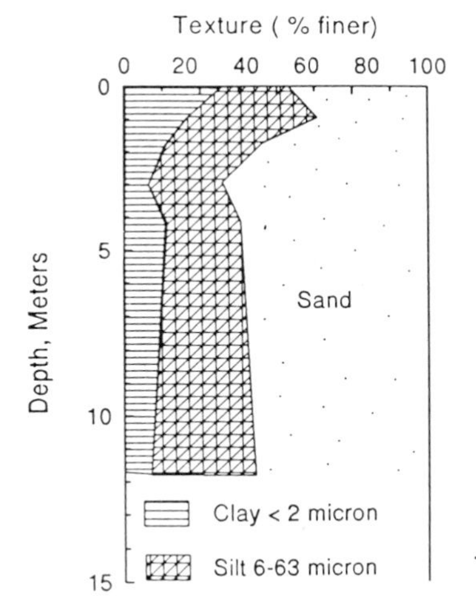

FIGURE 8.6. *Clay distribution of a soil-saprolite profile.* From Pavich et al., 1989, Plate 3. Republished by permission from J.J. Pavich.

Most studies show that the clay content is lowest in the saprolite and increases to a maximum in the Bt horizon (Fig. 8.6). The thickness of the saprolite zone, its texture, and chemical properties depend upon the type of bedrock (Fig. 8.3; Pavich, 1986; Pavich et al., 1989). For example, thick halloysitic saprolite occurs on felsic materials (Fig. 8.7), but mafic material has thin, mixed-mineralogy saprolite. Biotite and feldspars weather to kaolinite in felsic materials. Hornblende and other ferromagnesian minerals in mafic materials weather to smectite or kaolinite depending upon the environment. Dispersive materials beneath the Bt horizon are common where smectites form from mafic rock (Dadgari, 1982; Smith, 1986).

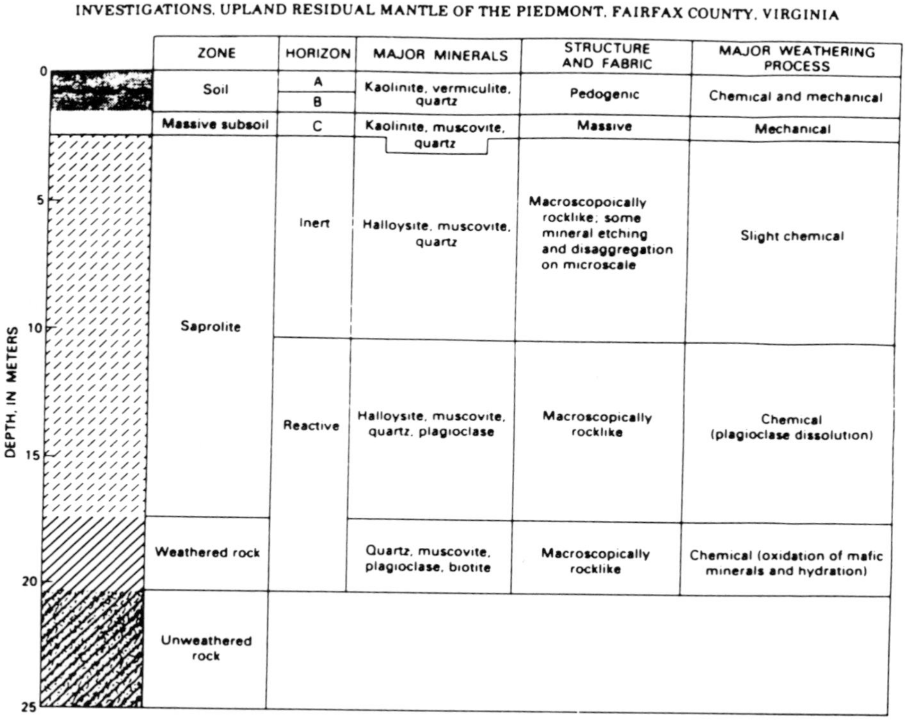

INVESTIGATIONS, UPLAND RESIDUAL MANTLE OF THE PIEDMONT, FAIRFAX COUNTY, VIRGINIA

ZONE	HORIZON	MAJOR MINERALS	STRUCTURE AND FABRIC	MAJOR WEATHERING PROCESS
Soil	A B	Kaolinite, vermiculite, quartz	Pedogenic	Chemical and mechanical
Massive subsoil	C	Kaolinite, muscovite, quartz	Massive	Mechanical
Saprolite	Inert	Halloysite, muscovite, quartz	Macroscopoically rocklike; some mineral etching and disaggregation on microscale	Slight chemical
	Reactive	Halloysite, muscovite, quartz, plagioclase	Macroscopically rocklike	Chemical (plagioclase dissolution)
Weathered rock		Quartz, muscovite, plagioclase, biotite	Macroscopically rocklike	Chemical (oxidation of mafic minerals and hydration)
Unweathered rock				

DEPTH, IN METERS: 0, 5, 10, 15, 20, 25

FIGURE 8.7. *Generalized weathering profile on quartzofeldspathic rocks.* Modified from Pavich et al., 1989, Fig. 17, p. 34. Republished by permission from M.J. Pavich.

In all but the coarse granites, the soils have clayey B horizons overlying loamy or sandy loam C horizons. Comminution of sand-sized pseudomorphs to clay-sized materials changes saprolite to a B horizon (Rebertus & Buol, 1986; Rebertus et al., 1986). Most of the mineral weathering occurs before the B horizon forms (Pavich, 1986; Rebertus & Buol, 1986; Rebertus et al., 1986). In the deeply weathered sections, the upper half of the saprolite is almost inert (Pavich et al., 1989).

Pavich (1986) believes there is an absolute loss of silica during B horizon development. The absence of a Si increase following volume reduction and increased density during B formation are the reasons for this conclusion. The Zr increase suggests the mass loss during transformation to soil is about 75%. Alteration of quartz and muscovite in the B horizon results from a large solution flux and long soil-zone residence. This suggests that the B horizon is slowly permeable. Other data show a high saturated conductivity only in the upper B horizon (Fig 8.8). Conductivity is low in the lower B, the BC and upper C horizons. The shape of the conductivity curves changes little with kind of bedrock.

During the growing season, there is little recharge to groundwater, but there is a rapid recharge during the autumn and winter. Shrinking and swelling of the clayey Bt horizon develops macropores for this rapid recharge (Pavich, 1986). Pavich (1986) estimates the volume change may approach 15% from air-dry to saturated conditions even in the kaolinitic clays of the Piedmont. The volume change probably is overestimated because most Bt horizons never become air-dry. The process probably is important because high clay contents are common in soils formed in felsic residuum.

Pavich (1986) constructed a chemical mass balance for the granite basin in Virginia. He concluded that dissolved solids in groundwater are from the lower part of the saprolite or the weathered zone. He suggested that water flux through the lower reactive zone is the primary control on rate of mass removal of dissolved solids.

The mass balance calculations show that plagioclase alteration is the major contributor to dissolved solids in the base flow. The high permeability of the weathered rock-zone at its interface with bedrock aids in the removal of dissolved solids. Much of the water movement in the saprolite is vertical, but bedrock joints may allow lateral flow.

Rate of Formation

From base-flow discharge, Pavich (1986) conservatively estimated saprolite formation at 1 cm in 3,000 years. He suggested that actual chemical mass loss

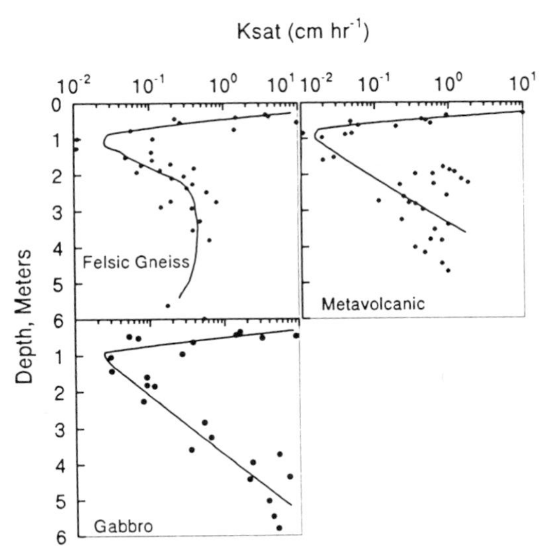

FIGURE 8.8. *Hydraulic conductivity of soil-saprolite profiles.* From Simpson, 1986, Fig. 3, p. 151.

may be greater by a factor of 2. Twelve m of saprolite form in 1.5 to 3 million years if the latter estimate is correct. Pavich et al. (1989) suggested that the average rate of saprolite formation is roughly 1 m/100,000 years. Thus, the oldest saprolite in the Piedmont may be no older than early Pleistocene. Mineral dissolution and saprolite formation are a continuing process and are most active at the bedrock interface. The age of the deposit, therefore, decreases with depth (Pavich et al., 1989).

Regolith thickness is the net balance of weathering and erosion rates (Pavich et al., 1989). This suggests that only the most stable parts of the landscape would have saprolite formed in early Pleistocene. Sloping Piedmont soils may have formed in materials considerably younger than early Pleistocene.

The age differences are of little value in soils studies where thick saprolites are present. Most mineral transformations take place at the rock/weathered rock interface or in the lower zone of the saprolite. Formation of a B horizon involves dissolution of silica but little transformation of other minerals. A possible exception is the formation of the ubiquitous 14-angstrom clay mineral in the near-surface horizons. The rate of formation and saprolite thickness vary with bedrock type. The processes leading to its development are the same, but the details may differ.

Rates of saprolite formation and subsequent thickness also may vary with landscape. Thicker saprolites should occur beneath the head and foot slopes where water concentrates, but few data are available. Most studies of thick saprolites are from tropical areas (Ruxton & Berry, 1957).

Several different kinds of residual deposits can form from weathering of surface materials, and we have touched only on a small part in the above discussion. For additional information and insight into the massive literature that exists we recommend Ollier (1969), Gardner et al. (1978), Wilson (1983), Colman and Dethier (1986), and Watson (1989) as good starting points.

REFERENCES

Becker, G.F. (1895). *A Reconnaissance of the Gold Fields of the Southern Appalachians.* U.S. Geol. Survey 16th Annual Rpt. 289–290.

Bricker, O.P., A.E. Godfrey, and E.T. Cleaves. (1968). Mineral-water interaction during the chemical weathering of silicates. In *Trace inorganics in Water* (pp. 128–142). Ed. by R.F. Gould. [Advances in Chemistry Series 73] Amer. Chem. Soc.

Cady, J.G. (1950). *Soil Sci. Soc. Amer. Proc.*, 15:337–342.

Calvert, C.S., S.W. Buol, and S.B. Weed. (1980). *Soil Sci. Soc. Am. J.*, 44:1096–1103.

Cleaves, E.T. (1974). *Petrologic and Chemical Investigation of Chemical Weathering in Mafic Rocks, Eastern Piedmont of Maryland.* Maryland Geol. Survey Rpt. of Invest. 25.

Cleaves, E.T. and J.E. Costa. (1974). *Equilibrium, Cyclicity and Problems of Scale—Maryland's Piedmont Landscape.* Maryland Geol. Survey Information Circ. 29.

Cleaves, E.T., D.W. Fisher, and O.P. Bricker. (1974). *Geol. Soc. Amer. Bull.*, 85:437-444.

Cleaves, E.T., A.E. Godfrey, and O.P. Bricker. (1970). *Geol. Soc. Am. Bull.*, 81:3015-3032.

Colman, S.M. and D.P. Dethier. (Eds.) (1986). *Rates of Chemical Weathering of Rocks and Minerals*. Orlando, FL: Academic Press.

Dadgari, F. (1982). *Pedogenesis of Na and Mg Affected Sedgefield Soils in the North Carolina Piedmont*. Unpublished doctoral dissertation. Raleigh: North Carolina State Univ.

Froelich, A.J. (1975). *Thickness of Overburden Map of Montgomery County, Maryland*. U.S. Geol. Survey Misc. Invest. Map I-920-B, scale 1:62,500.

Froelich, A.J. and T.L. Heironimus. (1977). *Thickness of Overburden Map of Fairfax County, Virginia*. U.S. Geol. Survey Open File Report 77-797, scale 1:48,000.

Gardner, L.R., I. Khorueromme, and H.S. Chen. (1978). *Geochimica et Cosmochimica Acta*, 42:417-424.

Hack, J.T. (1982). *Physiographic Divisions and Differential Uplift in the Piedmont and Blue Ridge*. U.S. Geol. Survey Prof. Paper 1265.

Langer, W.H. (1978). *Surface Materials Map of Fairfax County, Virginia*. U.S. Geol. Survey Open File Rpt. 78-78.

Nutter, L.J. and E.G. Otten. (1969). *Maryland Geol. Survey Rpt. Invest.*, 10:1-54.

O'Brien, E.L. and S.W. Buol. (1984). *Soil Sci. Soc. Am. J.*, 48:354-357.

Ollier, C. (1969). *Weathering*. New York: Elsevier.

Otten, E.G. (1981). The availability of ground water in western Montgomery County, Maryland. *Maryland Geol. Survey Rept. Invest.*, 34:38.

Pavich, M.J. (1986). Processes and rates of saprolite production and erosion on a foliated granitic rock of the Virginia Piedmont. In *Rates of Chemical Weathering of Rocks and Minerals* (pp. 551-590). Ed. by S.M. Colman and D.P. Dethier. Orlando, FL: Academic Press.

Pavich, M.J. and S.F. Obermeier. (1985). *Geol. Soc. Amer. Bull.*, 96:886-900.

Pavich, M.J., G.W. Leo, S.F. Obermeier, and J.R. Estabrook. (1989). Investigations of the characteristics, origin, and residence time of the upland residual mantle of the Piedmont of Fairfax County, Virginia. U.S. Geol. Survey Prof. Paper 1352.

Rebertus, R.A. and S.W. Buol. (1986). *Soil Sci. Soc. Am. J.*, 49:713-719.

Rebertus, R.A., S.B. Weed, and S.W. Buol. (1986). *Soil Sci. Soc. Am. J.*, 50:810-819.

Rice, T.J., Jr., S.W. Buol, and S.B. Weed. (1985). *Soil Sci. Soc. Am. J.*, 49:171-178.

Ruxton, B.P. and L. Berry. (1957). *Geol. Soc. Am. Bull.*, 68:1263-1292.

Simpson, G.G. (1986). *Soil Sci. Soc. of North Carolina Proc.*, 29:147-154.

Smith, C.W. (1986). The occurrence, distribution and properties of dispersive soil and saprolite formed over diabase and contact metamorphic rock in a Piedmont landscape in North Carolina. Unpublished master's thesis. Raleigh: North Carolina State University.

Watson, A. (1989). Desert crusts and rock varnish. In *Arid Zone Geomorphology* (pp. 25-55). Ed. by D.S.G. Thomas. New York: Halsted Press.

Wilson, R.C.L. (Ed.) (1983). *Residual Deposits*. [Geol. Soc. London Special Pub. no. 11] Oxford: Blackwell.

9 Geomorphology

INTRODUCTION

Geomorphology is the science of the earth's surface features: their character, origin, and evolution (Challinor, 1961). It also is "that branch of science that treats the surface features of the globe, their form, nature, origin and development, and the changes they are undergoing" (La Forge, 1925). Howell (1957) defined geomorphology as "the systematic examination of land forms and their interpretation as records of geologic history." Geomorphology and soil science should be a natural combination. The geomorphologists and pedologists need to know what processes shaped the landscape and when those processes started and stopped. Soils occur on landscapes and the past and present history of the site is of interest to a field pedologist.

In the past, geomorphology and soil science were largely descriptive. The emphasis was on landform or on where a soil with certain properties occurred. Both geomorphology and pedology now emphasize process. These fields are complex and require the knowledge of several other fields of earth sciences (hydrology, stratigraphy and geochemistry are examples).

GEOMORPHIC SURFACES

Definitions

A surface is a two-dimensional plane. It has width and length, but no thickness. The application of this definition to geomorphic surface makes data collection more rigorous, but it also results in better science.

A geomorphic surface is "a part of the land's surface defined in space and time" (Ruhe, 1969); "a landform or group of landforms that represent an episode of landscape development" (Balster & Parsons, 1968); "a part [of the earth's surface] that has been studied and mapped" (Daniels et al., 1971).

The surface of the earth is a composite of geomorphic surfaces that vary in age, shape and origin. A geomorphic surface can be level and planar, or steeply sloping and ranging from planar through concave and convex or any combination of those shapes. A geomorphic surface is an area of land that has a development history related in space and time. A geomorphic surface is either constructional or erosional or both. All erosional surfaces during their formation grade to a constructional surface.

Types of Surfaces

A depositional (constructional) surface is uneroded. It retains the shape left by the processes that deposited the underlying sediment. For example, an uneroded loess-mantled meandering stream deposit would have two constructional surfaces. The uneroded loess surface and the underlying fluvial surface are constructional features (Fig. 9.1).

An erosional surface is one formed by erosion and can have several shapes. It may range from nearly level to nearly vertical. Wind, mass movement and water develop erosion surfaces. Most erosional surfaces grade to depositional elements, even on slopes at 20%. The associated sediments, not slope, distinguish between erosional and depositional surfaces.

Constructional Surface Criteria Ruhe et al. (1967) established the following criteria for recognizing a constructional surface:

1. The complete weathering zonation typical of the area, underlies the surface.

   The Peorian loess in southwestern Iowa, for example, has a sequence of colors and calcareous and noncalcareous beds or zones (Fig. 9.2) that change properties as the loess thins toward the east. The gently convex to level stable interstream divides have the complete sequence of the local area and are part of the regional sequential change.

   The deeply weathered Pliocene to mid-Pleistocene terrace surfaces in the Atlantic Coastal Plain have weathering zones that contrast with those in the Iowa loess. Sediments with characteristic vertical zonation and high acidity underlie the well-drained North Carolina Upper and some Middle Coastal Plain divides (Figs. 9.2B and 9.3). Beneath younger and wetter surfaces, the properties of the weathering zones change in response to water-table levels and duration. The criteria defined by Ruhe (above) hold across a variety of sediments, ages and climates.
2. The surface parallels and does not truncate the weathering zonation (Fig. 9.3).

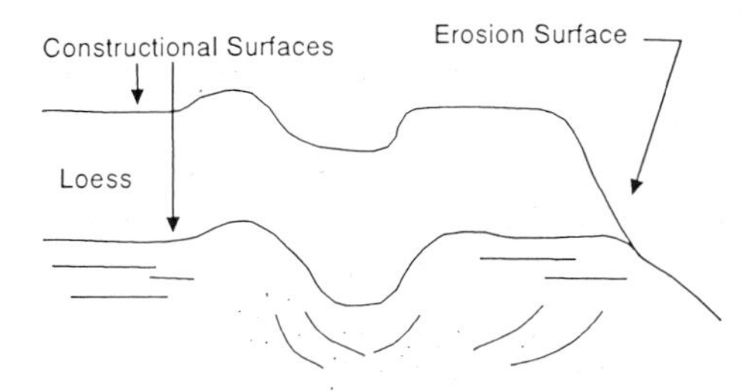

FIGURE 9.1. Depositional and erosional surfaces.

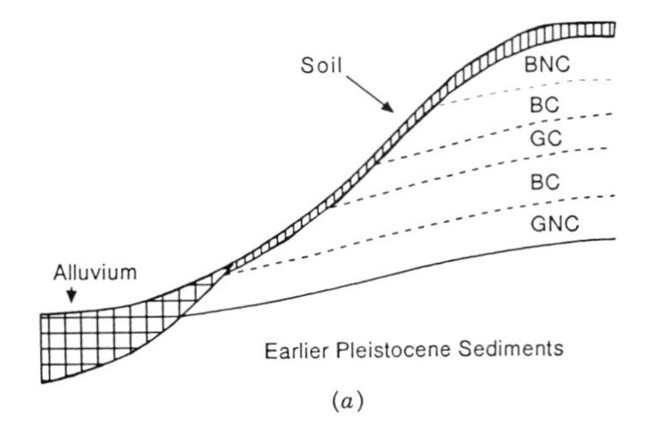

FIGURE 9.2. *Examples of regional weathering zonation*. A. BNC = brown noncalcareous. BC = brown calcareous. GC = gray calcareous. GNC = gray noncalcareous. Southwest Iowa. From Ruhe et al., 1955, Fig. 2, p. 346. B. Upper and Middle North Carolina Coastal Plain.

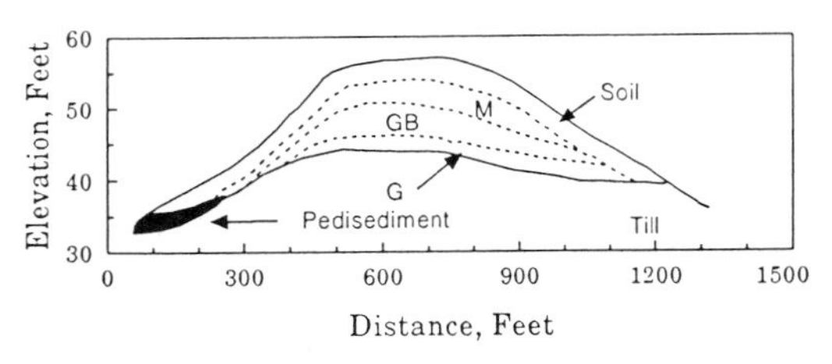

FIGURE 9.3. *Parallelism between interfluves and weathering zonation and truncation of weathering zonation by valley slopes*. From Ruhe et al., 1967, Fig. 69, p. 170. S = soil. M = mottled. GB = grayish brown. B = brown. G = gleyed.

This parallelism shows a close relationship in space and time between the surface and the weathering zones.

3. The surface parallels or closely parallels the base of the deposit.

   All erosion surfaces have some relief. Thus, only in a loess-mantled topography can one expect very close parallelism between the surface and deposit base.

4. The surface has a slope that is low enough to prevent erosion.

   Slope is a guide, not an absolute, in areas with dense vegetation and stable climate. Low slopes are no guarantee of stability in transition or tension zones where changes in kinds and density of vegetation are possible. For example, on nearly level surfaces with bunch-grass vegetation there is always the possibility of material moving downslope. The abundant debris dams and micro-channels with winnowing after a fire are evidence of erosion (Mitchell & Humphreys, 1987). Faunal activity exposes material to rainsplash erosion, even on gentle slopes.

5. The surface does not have a later deposit of local or major areal extent.
   A surface buried by eolian material may not be modified, but mass movement and water-deposited materials always have an erosion surface at the base.

Use the above criteria with an open mind; no criterion or combination of the above is absolute proof that erosion has not occurred.

Surface Properties

A geomorphic surface may be depositional, or erosional, or both in origin. It may be a level plain, a straight slope, or have a multicurvate or undulating shape. A surface may occur on one rock-type or sediment, or it may cut across several. It may have formed in a very short time (ash fall, lava flow), or have taken a long time to develop. A surface of large areal extent can have a considerable range in age. Tectonic movement, faulting or warping, do not destroy a geomorphic surface.

Many soil scientists have difficulty accepting the fact that a geomorphic surface has no thickness. Some pedologists also find it difficult to consider the surface as a separate entity from the underlying soil. For example, erosion that truncates the sediments and the overlying constructional surface destroys the surface in the truncated area. Erosion that removes a uniform thickness of material parallel to the surface also destroys the constructional surface. In either case, nearly all or none of the weathering zonation and original soil may remain. Erosion destroys the surface in that area. The soil profile is part of the weathering zonation, not the surface itself.

Not all workers agree that erosion destroys a surface. Some (Gile et al., 1981) define geomorphic surfaces by genetic landform, geologic age, and related pedogenic features. They allow some erosion and base their identification more on sediment and soil properties, including weathering zonation. They maintain that the weathering profile is part of the expression of the geomorphic surface. Their surface has three, not two, dimensions. We agree that the weathering zonation and soils are part of the geomorphic history, but so is the later erosion. We lose a part of the site history by not recognizing the later erosion. It is easy to recognize that some properties of the sediment or soil are from earlier weathering.

Dating Geomorphic Surfaces

A two-dimensional surface has properties that allow us to give it a relative age or an absolute age if datable material is present (Fig. 9.4). A geomorphic surface is younger than the youngest deposit or land surface it cuts. It is younger than a higher surface to which it ascends (Ruhe, 1969). The ascendency and descendency law states a hillslope is the *same age* as the sediment to which it descends. A hillslope is *younger* than the higher surface to which it ascends.

Trowbridge's criterion (1921) for erosion surfaces states that

1. *An erosion surface is younger than the youngest material it cuts.* Surface II in Figure 9.4 is younger than Pliocene sediment 2.
2. *An erosion surface is younger than any structure it bevels.* Surface II is younger than fault 8, and surface IIIA is younger than fault 9.
3. *An erosion surface is younger than materials forming erosional remnants above it.* Surface IV is younger than the erosional remnant of sediment 3.
4. *An erosion surface is younger than any adjacent erosion surface that stands at a higher level.* In Figure 9.4, surface III is younger than II or IIa.

An erosion surface is

1. *Contemporaneous with the overlying alluvial deposits.* Erosion surfaces IIa and IIIa are the same age as the immediately overlying beds in sediments 4 and 5 (Fig. 9.4). The erosion surface at the top of sediment 1 is the same age as the adjacent beds of sediment 2.
2. *The same age or older than other overlying terrestrial deposits.* If sediment 3 overlies an erosion surface on sediment 2, then this erosion surface is the same age as sediment 3. Only eolian sediments can overlie an older surface. Nearly all marine and fluvial sediments have an erosional base. Thus, the contact with the older sediment is the same age as the overlying younger material.

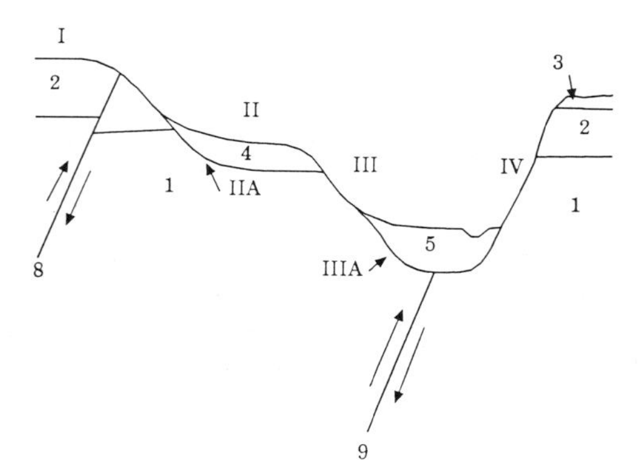

FIGURE 9.4. *Illustrating ages of surfaces on a landscape using Trowbridge's criteria.* 1. Cretaceous. 2. Pliocene. 3. Mid-Pleistocene. 4. Late Pleistocene. 5. Holocene. Faults: 8. 2,000,000 yrs. 9. 9,000 yrs.

3. *Older than valleys cut below it.* Surface II is older than the valley that contains surfaces III and IV.
4. *Older than deposits in valleys below it.* Surface II is older than sediment 5. There are exceptions to the above (4) rule. For example, an erosion surface on the valley slopes formed by mass movement can be younger than the lower valley deposits.
5. *Older than any lower adjacent erosion surface.* Surface II is older than surfaces IIIa and IV because they are in the lower valley.

A constructional surface is

1. *The same age as the immediately underlying sediment.* In Figure 9.4, surface I and constructional surfaces II, III, and IV are the same age as the underlying sediments.
2. *The same age as the erosion surface to which it ascends in a smooth concave upward profile.* Constructional surfaces II, III and IV are the same age as their higher erosional counterparts. Each surface has a smooth transition from erosional to depositional origin.

Recognition of constructional surfaces is moderately easy on broad plains. Proof usually requires considerable field work, especially when the surfaces are of minor extent in dissected landscapes.

Major erosion surfaces occur in many parts of the world, but the subaerial exposure is small. Erosion surfaces of moderate to low slope usually have a mantle of hillslope sediment or pedisediment of Ruhe et al. (1967) and Ruhe and Walker (1968) derived from upslope. Pediments are another example of surfaces with little of the erosion surface exposed although the areal extent beneath a mantle of debris may be large.

HILLSLOPE NOMENCLATURE

Geomorphologists have studied hillslopes since the science started. The result is an abundant literature describing various ideas of how hillslopes evolve. There also is an extensive literature on hillslope nomenclature.

For our purposes, we will use the simple and easily understood nomenclature proposed by Ruhe (Fig. 9.5). The advantage of Ruhe's system is that it helps one place a site in relation to other elements of that landscape. Ruhe devised the system for landscapes with both erosional and constructional surfaces, so be cautious in sloping landscapes with narrow ridges.

Using Ruhe's terminology in depositional systems does not always convey the proper meaning to those working in erosional landscapes. Do not use the term *summit* where coalescing backslopes have narrowed the interfluve or finger ridges to the point that they are now *downwearing*. Under these conditions, the summit disappears and only a shoulder remains. Another exam-

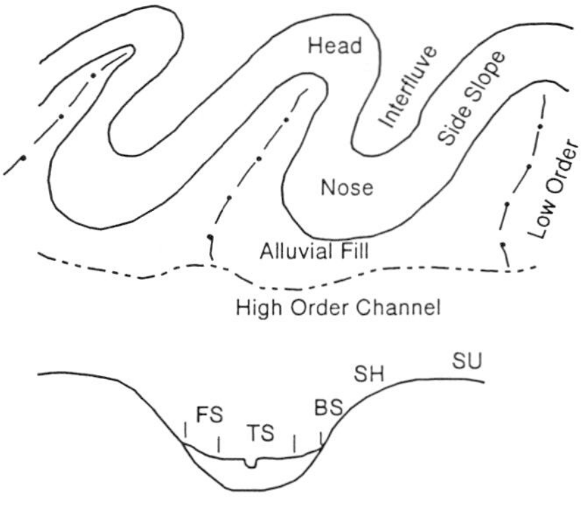

FIGURE 9.5. *Landscape elements*. SU = summit. SH = shoulder. BS = backslope. TS = toeslope. FS = footslope. From Ruhe, 1975, Fig. 6.2, p. 101. Reprinted by permission from R.V. Ruhe.

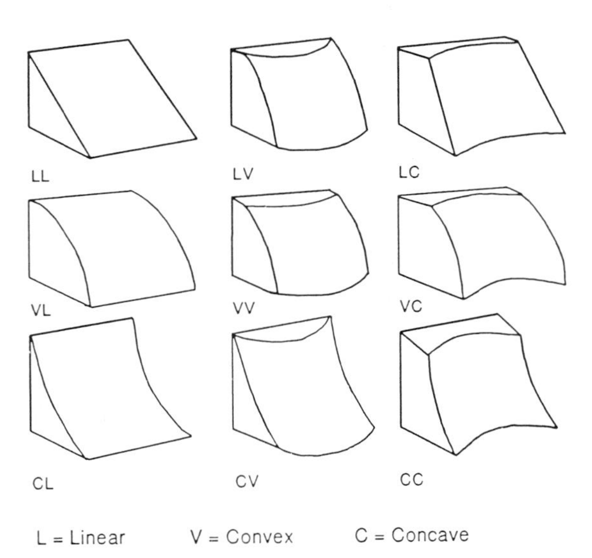

FIGURE 9.6. *Geometric forms of hillslopes*. L = linear. V = convex. C = concave. The first letter refers to the downslope shape and the second to the cross-slope shape. From Ruhe, 1975, Fig. 6.1, p. 100. Reprinted by permission from R.V. Ruhe.

ple is dunal landscapes, where a slope may have a different genesis depending upon whether it is on the windward or lee-side.

There are many ways to quantify site shape and location. The methods range from geometric description of hillslopes (Fig. 9.6) to highly detailed contour maps (O'Loughlin, 1986; Sinai et al., 1981). Each method has its place, but Ruhe's nomenclature is useful for locating oneself on many landscapes. It helps one predict soil and sediment properties (Figs. 9.5 and 9.6) because hillslope sediments change property's downslope. Other methods may be more quantitative, but none is more helpful during the initial phases of field work.

REFERENCES

Balster, C.A. and R.B. Parsons. (1968). *Geomorphology and soils, Willamette Valley, Oregon*. Oregon State Univ. Agri. Exp. Sta. Spec. Rpt. 265.

Challinor, J. (1961). *A Dictionary of Geology*. Cardiff: Univ. of Wales Press.

Daniels, R.B., E.E. Gamble, and J.G. Cady. (1971). *Advances in Agron*., 23:51–88.

Gile, L.H., J.W. Hawley, and R.B. Grossman. (1981). *Soils and Geomorphology in the Basin and Range area of Southern New Mexico—Guidebook to the Desert Project*. New Mexico Bur. Mines and Min. Res., Memoir 39.

Howell, J.V. (1957). *Glossary of Geology and Related Sciences*. Amer. Geogr. Inst.

LaForge, L. (1925). *Geol. Survey Georgia Bull*., 42.

Mitchell, B.P. and G.S. Humphreys. (1987). *Geoderma*, 39:331–357.

O'Loughlin, E.M. (1986). *Water Resources Res*., 22:794–804.

Ruhe, R.V. (1969). *Quaternary Landscapes in Iowa*. Ames: Iowa State Univ. Press.

Ruhe, R.V. (1975). *Geomorphology*. Boston: Houghton Mifflin.

Ruhe, R.V., R.B. Daniels, and J.G. Cady. (1967). *Landscape Evolution and Soil Formation in Southwestern Iowa*. USDA Tech. Bull. 1349.

Ruhe, R.V., R.C. Prill, and F.F. Riecken. (1955). *Soil Sci. Soc. Am. Prof*., 19:345–347.

Ruhe, R.V. and P.H. Walker. (1968). *Trans. 9th Int. Cong. Soil Sci*., 4:551–560.

Sinai, G., C. Zaslavsky, and P. Golany. (1981). *Soil Sci*., 132:367–375.

Trowbridge, A.C. (1921). *University of Iowa Study in Natural History*, 9:7–127.

10 How Landscapes Evolve

INTRODUCTION

Almost all field scientists need to have a working knowledge of how landscapes evolve through space and time. The reason for this statement is that the processes responsible for the landsurface left behind materials, older surfaces and truncated weathering zones and shapes that control the flow of energy and water to and from each component. Any field scientist recognizes the spatial variability of almost all landsurfaces and the various components it supports (soils, ecosystems and hydrologic systems, for example). The major question is how does that variability affect the particular system of interest? Also, how far can the data from one component of the landscape be extrapolated to other similar and dissimilar landscapes?

Almost all problems related to water quality and availability, waste disposal, land reclamation, the impact of global warming on terrestrial ecosystems and environmental quality require a knowledge of the landscape and its soils. The development of landscape ecology is partially a result of the focus of these above issues on the landscape and soils. The treatment given soils and geomorphology by landscape ecologists is rather superficial (see Forman & Gordon, 1986; Naveh & Lieberman, 1984). Rowe (1984) clearly defined the problem when he stated that the progressive narrowing of scientific disciplines leads to scientists studying the parts before the systems are recognized.

THEORIES OF LANDSCAPE EVOLUTION

Landscapes evolve so slowly that it is difficult to quantify changes that occur. Man-induced erosion in cultivated or other disturbed areas compounds the challenge. In severely disturbed areas, it is difficult to piece together the relationships among soils, sediments, and geomorphology into an approximation of what once existed. Deep gullying has removed materials related to the preexisting landscape, and sedimentation locally has buried the preexisting soils and sediments. We therefore cannot satisfactorily test ideas on how many landscapes have evolved. As a result of all the above, the literature contains several different opinions about how landscapes evolve. Yet it is important in field work to have some idea of how landscapes change over time naturally and as affected by man. We also need to know what possible mechanisms produce these changes. Different processes can produce the same surface form, so we must remember that it is possible to get to the same point from several different directions.

A massive literature details several theories on how landscapes evolve. We will discuss only five, however. They are Davis's peneplanation (1899), Penck's parallel retreat (1953), Hack's dynamic equilibrium (1960), Ruhe's ideas of backwearing (Ruhe et al., 1967) and Dalrymple and associates' (1968) nine-unit landsurface. Some field evidence supports any one of the five if one ignores other evidence. The landscapes studied by each particular writer may have had a major influence upon what ideas developed.

We suspect that landscapes within a given environment have developed from several processes. Also, the morphological expression of these processes may vary considerably. We need to have some idea of how landscapes evolve, but it is a mistake to embrace only one model. In other words, we must all keep testing ideas and be willing to change as new data warrants change.

PENEPLANATION

Davis (1899) had a tremendous influence on ideas of landscape evolution from the late 1800s through the early part of this century. He was a prolific and convincing writer, and his ideas of peneplanation resulted in the identification of several peneplains.

Figure 10.1 illustrates Davis's ideas of how a landscape evolved in a humid climate. Uplift of a level landmass at time 0 reaches its maximum elevation at time 1. Point B represents the average altitude of the high parts, and A represents the average altitude of the low parts. By time 2, the large rivers had reduced their altitude to C while the divides lowered somewhat to D. Initially, the relief increases as the riverbeds continue to lower toward base-level. Water does not concentrate on the divides, and they weather away slowly at this stage. From time 2 to 3, the relief increases to a maximum as steep-sided valleys deepen and side-valleys extend headward.

At time 3, the divides are still being lowered. Between times 3 and 4, the reduction of relief is at a maximum, and the slope of the valley sides is decreasing. The changes from time 3 onward are slow, but the reduction of relief con-

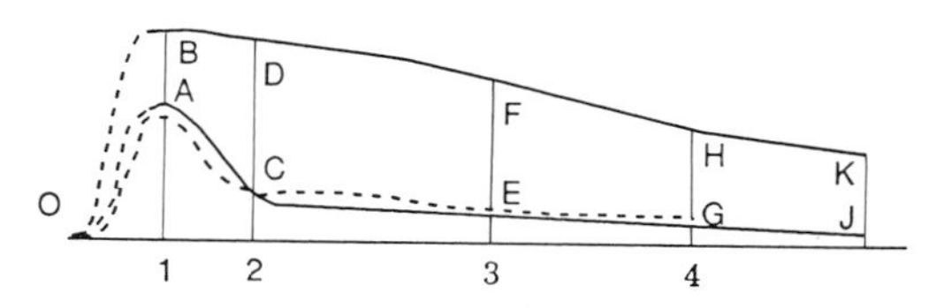

FIGURE 10.1. *Landscape evolution by peneplanation.* Redrawn from Davis, Fig. 4, p. 255, 1954.

tinues until only a rolling lowland exists. Davis believed that if enough time elapsed, these changes occurred regardless of the original height. The result of the cycle is a *peneplain* (*pene* = almost), or a very gently sloping landscape. All parts of the peneplain graded to the adjacent stream channel.

One idea Davis used in developing his peneplain hypothesis is that valley floors do not reduce to absolute base-level. The altitude of any point on a well-matured valley floor depends on the river slope and distance from the river mouth. Hack (1957) has shown that the above is true, but he measured from the stream's divide. Erosion forces the initial flat crests or broad swells to migrate, so strong structural belts maintain the subsequent divides. This means an increasing adjustment of both streams and the landscape to bedrock structure. Davis also believed that rivers eroded their valleys, but surface-wash removes major amounts of debris. Freezing and thawing, chemical disintegration, solution, creep and other mass-wastage mobilizes materials.

Davis stated that when uplift ended none of the original surface, divides and slopes, would remain. Throughout the cycle, there is continual removal of material from the surface. This means that relict landscapes do not exist. Low erosion rates on the divides compared to the valley slopes allows deep weathering on the divides. The divide surface lowers constantly.

PARALLEL SLOPE RETREAT

Walther Penck's (1953) ideas on how landscapes change are in sharp contrast to those of Davis. The fundamental difference is that Penck explained slope evolution under unchanging conditions as one of uniform removal. Under these conditions, the slope retreats parallel to itself above a more gentle consuming slope (Fig. 10.2). Weathering of planer slope t-1 starts at t0. With time, uniform removal of a wedge of material at the base results in a gentler slope at the base, t-2. Weathering removes the same wedge of material along the face above the base. The face retains the original declivity (Fig. 10.2, 2′-2).

Continued uniform weathering and debris removal eventually result in slope t-t′ that consumes the original slope. Throughout the period, the original slope retreated parallel to itself, so the upland divide did not change shape or elevation. The divide was stable throughout that period unless a slope encroached from the other side. This idea of backwearing under uniform conditions allows relict landscapes to exist, and several geomorphic surfaces can occur within one watershed.

Penck has several ideas that are worth considering when studying landscapes.

- *Principle of flattening.* Denudation (removal of rock at the same rate of rock reduction) flattens a slope; it never steepens it.
- *Development of slopes.* Vigorous erosion develops convex or straight slope profiles, and waning erosion develops concave slopes.

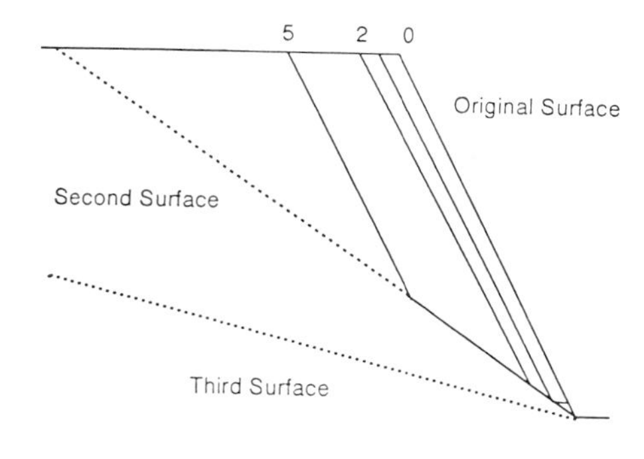

FIGURE 10.2. *Penck's parallel retreat of a rock face.* Redrawn from Penck, 1953, Fig. 2, p. 135.

- *Flattening of slopes.* A cliff rising from the edge of a non-eroding stream retreats by denudation. A gentler basal slope develops below it, and an even gentler slope develops below the basal slope.
- *Straight slope profiles.* Slopes move further from the stream edge if the base-level of erosion remains unchanged. Eventually the upper slope unit or units disappear. The lower slope unit continues to grow in size.
- *Convex breaks in slope.* An increasing intensity of erosion produces a convex break in slope. The convexity starts at the stream and works upward at a rate depending on the intensity of denudation of the steep lower unit. The higher, flatter slope development depends upon the position of the break in slope. The higher slope becomes independent of the base-levels of erosion.
- *Continuity of curvature of slopes.* A steeper slope unit overtakes a continuously curved slope. A markedly curved slope develops that recedes upward as a convex break in gradient. This convex break in slope marks a discontinuity in the slope units.

Penck's ideas on how landscapes can develop strongly suggest those relict landsurfaces remain unmodified until consumed from below. This idea allows surfaces unmodified by erosion on the higher parts of the landscape away from the influence of the stream. These surfaces are independent from erosional activity on the adjacent valley slopes. His ideas suggest that landscapes above a convex break in slope can considerably predate the younger bounding slopes. Old landsurfaces can occur next to young landsurfaces. Davis's ideas say that all parts of the landscape are about the same age (surface). He believes the deep weathering is only the result of slow removal, not stability.

DYNAMIC EQUILIBRIUM

John Hack (1960) advanced the idea of dynamic equilibrium as a model for landscape development. He assumes mutual adjustments of all topographic elements so they are *downwasting at the same rate.* The rate can vary considerably depending upon the material. The forms and processes are in a steady state of balance and are time independent.

There would be no relict landforms or landscapes by the above definition, but Hack recognizes relict forms such as terraces and coastal plains. Dynamic equilibrium requires a state of balance between opposing forces. These forces operate at equal rates and effectively cancel each other to produce a steady state. This means that an alluvial fan is losing as much material as is being deposited. If one takes the narrow view of Hack's ideas, then once established, a landscape would change little in form. The areal extent of the various units and their placement within the landscape would change over time, however.

The idea of dynamic equilibrium has wide application for systems with equal inflow and outflow of energy. For example, the flow of groundwater and surface water adjusts to the landscape. Additionally, groundwater flow changes with each landscape modification. A level, permeable surface sediment overlying a confining bed may have little water moving through it before landscape dissection. The underlying confining layer may keep the surface material saturated most of the year if rainfall exceeds evapotranspiration. A thick blanket of peat forms above the mineral surface if surface run-off and evapotranspiration are less than rainfall. Even under these conditions, there is a balance between the influx and outflux of energy.

After deep dissection by the stream system, elements of the original surface remain. Water can now move through the sediment to outlets adjacent to the stream channels or to seeps on the valley slopes. Intensive weathering of minerals can occur within the sediment because leaching removes the weathering products. Previously, most of the weathering, if it occurred, was in the very upper part of the sediment. Patches of peat may persist in a few areas. However, a much different equilibrium is being established, even though the influx of energy and water is the same.

BACKWEARING

Ruhe's (Ruhe et al., 1967) backwearing model developed from extensive field work on properties of the late Sangamon pediment in southwest Iowa (Fig. 10.3). His work shows that stream trenching initiates the development of an extensive erosion surface. From an initial nearly level surface (Fig. 10.3A) the controlling stream and tributary channels incise. Erosion proceeds from the stream into the upland. The initial channel trenching leaves deep, steeply sloping walls. These are the initial valley slopes. These steep slopes or walls erode and start to grade to the adjacent stream channel in a concave upward profile. This requires the upper part of the slope to recede and lose more material than the area next to the channel. The shoulder between the uneroded upland and the valley slopes recedes from the channel at b toward b′ (Fig. 10.3A, B). Streams in pseudoequilibrium have a concave upward profile

(Hack, 1957), or decreasing slope downstream. Valley slopes formed by water erosion develop the same profile (Meyer & Kramer, 1969). For any depth of incision under constant conditions, there is a maximum length that the channels can deepen upslope. In Ruhe's example (Fig. 10.3), erosion will cease where the upslope increase in channel or erosional gradient cuts the original surface.

Later work (Schumm et al., 1987) shows that depth of trenching from a lowered base-level decreases upstream. Schumm's work supports Ruhe's idea of decreasing depth of dissection upstream or upslope. The resulting landscape has preserved remnants of the old landsurface. A younger pediment slopes toward the valley floor and merges almost imperceptibly with the side-valley and flood-plain alluvium. A lag develops if the material has gravel that erosion cannot move. A thin veneer of sediment (hillslope sediment) that usually overlies the pediment buries the lag. Ruhe's model works well in many landscapes, both in and out of the glaciated regions. His ideas are well worth testing when one is trying to develop the landscape history of an area.

NINE-ELEMENT LANDSCAPE

Conacher and Dalrymple (1977) developed a nine-element landsurface unit (Fig. 10.4) that emphasizes the pedogenic processes and their relative intensity operating on various parts of a landscape. Most landsurface elements are present in erosional landscapes. This model graphically illustrates the relationship of each part of the landscape to the other, especially down the hydraulic slope. The nine-element landscape model also emphasizes proc-

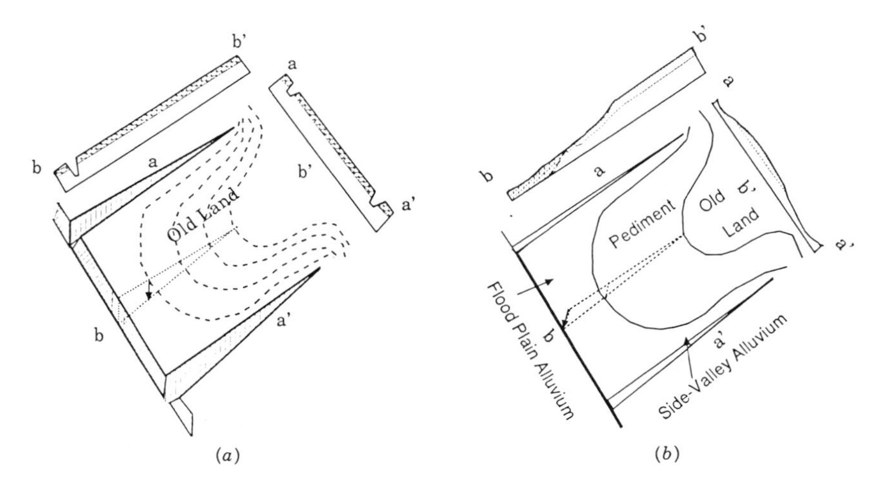

FIGURE 10.3. *Evolution of the Late Sangamon pediment, fan and flood plain*. From Ruhe et al., 1967, Fig. 51 A, B, p. 134.

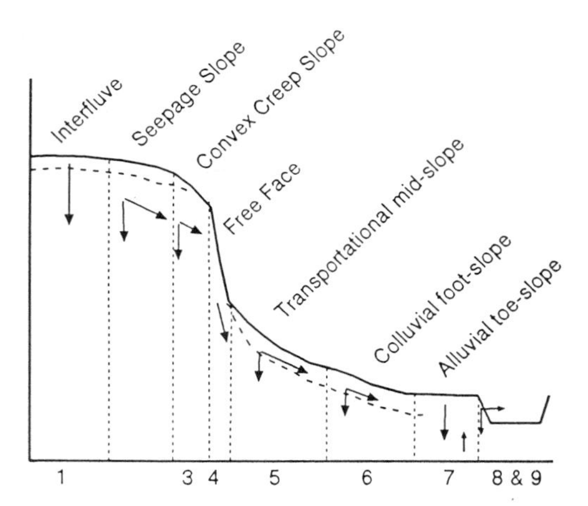

FIGURE 10.4. *Nine-unit slope model.* Redrawn from Dalrymple et al., 1968, Fig. 1, p. 61. Reprinted by permission from Gebruder Borntraeger.

esses that are active on each segment. The different processes indicate why soil properties vary depending upon landscape location.

The model considers only the process responsible for various parts of the landscape. Although landscape processes are important in understanding soils, the relative or absolute ages of parts of the landscape also are important in understanding the areal extent and properties of soil.

APPLYING LANDSCAPE MODELS TO FIELD RESEARCH

Most landscapes evolve too slowly for reasonable testing of ideas on how they change. However, it is important for field scientists to have some knowledge of the subject. Field work involves a constant testing of ideas; if one's predictions are good, one's ideas may have some validity in that area. They may be far from the truth, however. When predictions fail, prepare to discard or revise the ideas.

Each of the landscape models given above probably will work well in some landscapes, or in parts of a landscape. For example, the southeastern Piedmont is a gently to steeply rolling landscape. All parts of the local landscape seemingly grade to the adjacent ephemeral or perennial stream channel. The shape of the landscape and its often gentle relief seem to fit Davis's ideas of a peneplain.

The Coastal Plain landscape is in sharp contrast to the Piedmont. The Coastal Plain has broad, gently undulating flats with many depressions and sharp boundaries to the later valley slopes. In the Coastal Plain, Davis's ideas do not appear to explain the observed relationships as well as those of Penck and Ruhe. In the Coastal Plain, a combination of Ruhe's model and

Hack's dynamic equilibrium appears to explain most of the soil-geomorphic relationships.

Let us consider these models with a third example, the Cumberland Plateau. The folded sandstone and shale of the Cumberland Plateau make up a distinctive landscape. The sandstone is more resistant to weathering than shale, and it caps most of the ridges. The shapes and width of the ridges are a function of the local folding of the sandstone. Colluvium with some sandstone fragments mantles shale side-slopes. Steep slopes are next to major tributaries, but gentle slopes grade to the first order channels in the headwaters. The most resistant rock controls the base-level of major streams in landscapes with folded materials of different weathering resistant. However, the least resistant rock controls stream placement and valley slope location.

What landscape development model best fits the Cumberland Plateau area? One could probably find parts of the landscape that would correlate well with any one of the models discussed. However, the process models such as Hack, Ruhe or Dalrymple probably are more useful in helping explain soils than the others.

EVOLUTION OF A THEORETICAL LANDSCAPE

We will use Ruhe's model to illustrate the various stages of development of a landscape. The initial landscape has a nearly level surface with some surface drainage. The underlying materials are uniform, so no landform is the result of bedrock hardness. Streams deepen their channels in response to a lowered base-level, but little trenching occurs after the initial deepening. The base-level rises slightly after the trenching in response to alluviation. We will outline the development of the controlling stream and ephemeral channel system and their effect on the upland following two periods of channel deepening. The purpose of this exercise is to illustrate how a landscape may evolve in sequential steps and what morphological expressions of the processes remain. Landscapes can evolve from several different processes; we are only trying to illustrate the simplest evolution related to surface erosion from stream trenching.

Even nearly level surfaces have low areas where water collects. The initial development of drainage on a nearly level surface will follow the natural lows of the surface (Fig. 10.5A) where water volume and unit stream power are at a maximum (Moore & Burch, 1986). This combination results in a potential for maximum erosion in the natural low, especially during trenching of the major stream. The nontrenched thalwegs shown in Figure 10.5A are common in the nearly level Coastal Plain surfaces of the southeastern United States. These areas are only slightly lower than the surrounding landscape, and a definite channel does not exist.

The local base-level lowers 10 m by trenching of the main channel. The tributary channels also trench and work headward into the upland along the nontrenched thalwegs of Figure 10.5A. A 10 m deepening of the main channel

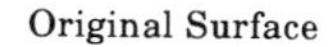

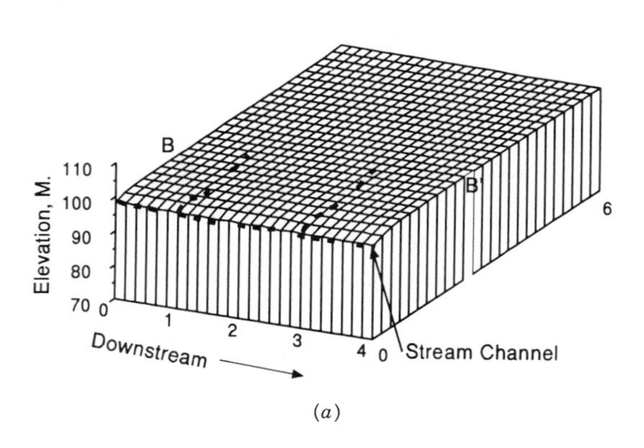

(*a*)

First Dissection Landscape

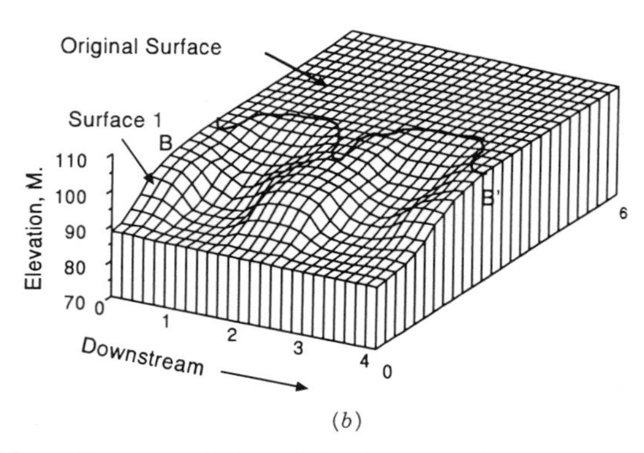

(*b*)

FIGURE 10.5. *Three-dimensional view of development of a multi-incised landscape from an original nearly level surface.* A. Initial landscape. B. Areal extent of original surface and erosion surface 1 at end of first erosion cycle. C. Areal extent of original surface and erosion surfaces 1 and 2 after second stream incision and partial adjustment of valley slopes.

does not mean the same depth of cutting throughout its tributaries (Fig. 10.6). The incised tributary channel gradients increase upstream at a rate greater than the upland slope. Thus, the depth of trenching of these tributaries decreases upstream. The trenching ends in a convex reach that merges with the nontrenched original channel (Fig. 10.6A).

The initial dissection affects only the area near the main and tributary channels. The channel walls have high declivity and a sharp break from the nearly level, uneroded original surface (Fig. 10.6). Most of the landscape is unaffected by channel trenching at this stage.

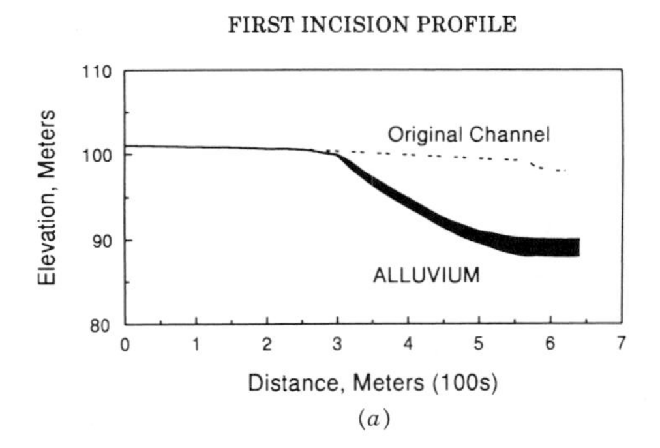

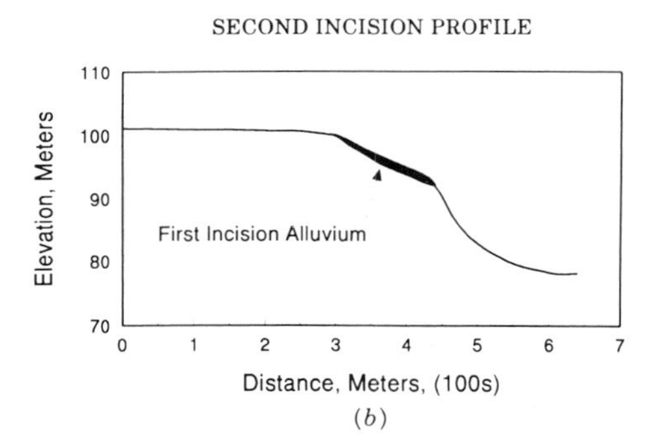

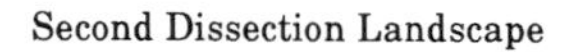

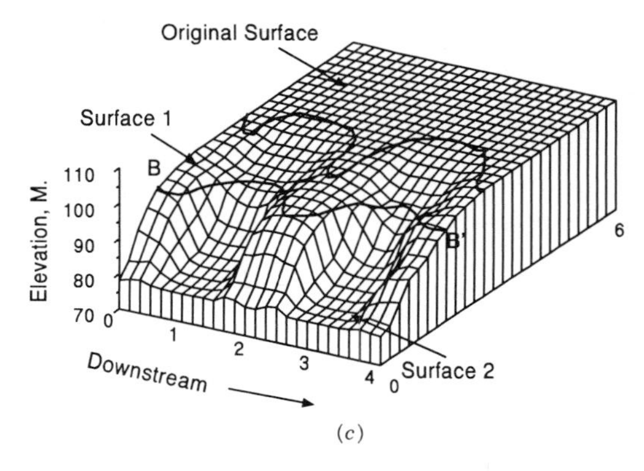

FIGURE 10.6. A. Profile of thalweg at grid point 1 (Fig. 10.5A) immediately after main channel trenching (dashed lines) and after headward extension into the uplands and alluviation. B. Profile after second incision of main channel.

The steep channel walls start to erode to the new base-level after the tributaries deepen. Although the potential for maximum erosion is at the channel, this is also the area of maximum deposition after the initial trenching. Meyer and Kramer (1969) simulated the development of slopes grading to a stable base-level using the Universal Soil Loss Equation (USLE) as the model. They found that all downslope profiles—linear, complex, or convex—develop a convex-concave profile. The zone of maximum erosion is near the midslope, and their data parallel that of Moore et al. (1986).

After the initial trenching, the steep channel walls retreat toward the local divide, and their slope gradient decreases. During slope retreat for one base-level, the upper slopes recede toward the local drainage divide more than those slopes next to the channel. Sediment deposition in the channel buries the erosion surface and develops a convex-concave shape beneath the alluvium (Fig. 10.7). A thin sediment deposit blankets the lower slopes, and the valley slopes merge with the valley-fill in a smooth concave profile. The slopes above the alluvium are linear to slightly convex. The slopes above the thin mantle of alluvium are younger, eroded later, than the erosion surface below the alluvium.

Eventually valley slopes of moderate to low gradient merge and the channel is a smooth concave profile (Figs. 10.5B, 10.6A and 10.7A). Erosion attacks the upland normal to the tributaries, and locally lowers the narrow interfluve (Figs. 10.8 and 10.5B). The interstream or interfluve divide lowers when opposing tributaries narrow the divide to a critical, but unknown, width. The new

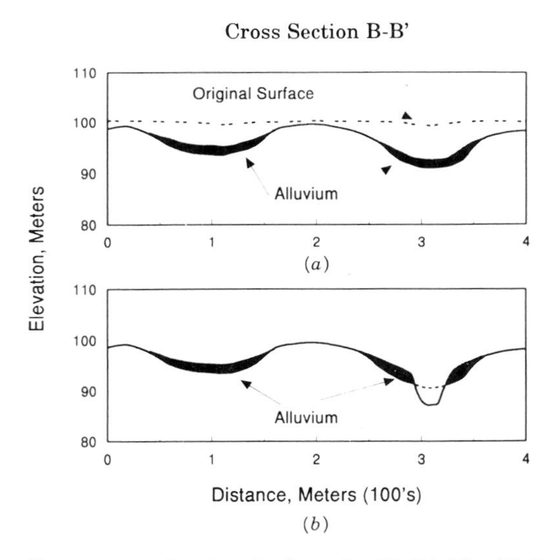

FIGURE 10.7. *Profiles across upland and tributaries* (B–B′, Fig. 10.5). A. Original and first incision profile after alluviation. B. Profile after second incision.

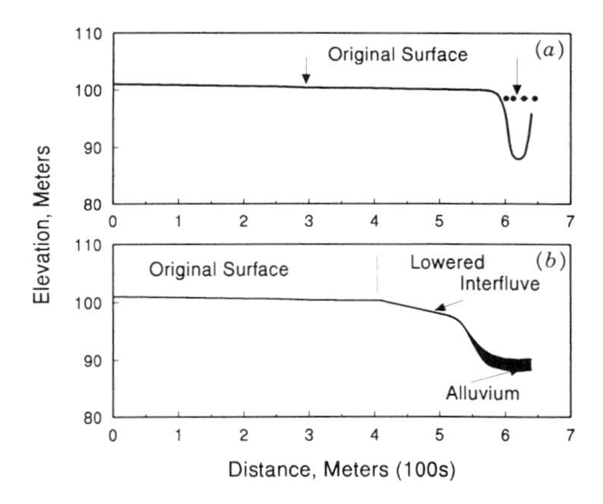

FIGURE 10.8. Profiles of landscape from interstream divide to main channel at grid point 2, Figure 10.5, showing (A) initial surface, solid and dashed line, and (B) profile at end of the first incision erosion.

erosion surface destroys a large part, *but not all*, of the initial surface (Figs. 10.5B, 10.6 and 10.8).

Erosion always means deposition somewhere, either nearby or as far away as the ocean. As the landscape develops, alluviation occurs in the main valley and its tributaries (Figs. 10.6, 10.7 and 10.8). Geology and soil maps of suitable scale recognize these deposits. However, these maps usually ignore the abundant but thin veneer of hillslope sediment that mantles the lower slopes (Fig. 10.6).

Later stream incision renews the process, with erosion removing some of the landscape of erosion surface 1 (Figs. 10.5C, 10.6B and 10.7B). The interstream divides, and parts of the first incision landscape (Fig. 10.5C) remain intact. Each surface grades to a former or present stream channel (base-level) responsible for its formation (see Fig. 10.6B).

The development of the upland landscape follows the profiles found in natural streams. Streams in pseudoequilibrium have a decreasing slope downstream in response to the increase in run-off and watershed area (Hack, 1957). Valley slopes formed by water erosion have the same profile (Meyer & Kramer, 1969).

Stream channels extend headward into the upland following a lowered base-level. For any decrease in base-level, there is a maximum distance the headcut can move upchannel (Schumm et al., 1987). The maximum depth of erosion is at or near the main channel and decreases progressively upstream (Fig. 10.7). Multiple lowering of the base-level results in a stepped sequence of channels and stabilized headcuts throughout the tributary (Table 10.1). Channel slopes of the later incision are steeper than the former channel the incision replaced (Schumm et al., 1987). Part of the adjacent valley slopes grade to the new base-level. This results in one or more sets of valley slopes grading to

TABLE 10.1. Slope, Correlation Coefficients and Meters above Present Stream for Two Ephemeral Channels in Guilford County, NC

Reach	b value	r	Significant Level	Meters above Controlling Channel
		Thalweg 1		
1 Surface^{+}	−0.273	−0.772	ns	
2 Surface	−7.699	−0.974	***	5.6
3 Surface	−17.264	−0.992	***	3.5
		Thalweg 2		
1 Surface	−1.565	−0.801	**	16.5
2 Surface	−22.637	−0.993	***	10.2
3 Surface	−25.619	−0.967	**	8.7

$^{+}$Surface of hillslope sediment or alluvium.
ns = Not significant.
** = Highly significant 0.01 level.
*** = Very highly significant >0.01 level or 0.001.

From McCracken et al. (1989), Table 2, p. 1148. Reprinted by permission from Soil Science Society of America.

channels developed under a higher base-level than present (Figs. 10.5C and 10.7B). These slopes and channel systems occur throughout the watershed.

The landscape resulting from one or more channel incisions usually will include remnants of one or more old landsurfaces. The old land and its ephemeral channels grade toward the present main valley. These older remnants, however, grade to channels considerably above the modern valley floor. They grade into thin air. A younger valley slope, or sequence of slopes, grades to the adjacent valley floor (Figs. 10.6 and 10.7). All slopes of the same unit grade to an adjacent fill, or a younger unit truncates them. Only the youngest slopes grade to the modern flood-plain alluvium.

The above scenario closely fits the Piedmont landscape and many landscapes seen in other areas. The Ruhe model, the basis for the theoretical landscape shown in Figures 10.5 to 10.8, is useful in areas with many kinds of soil materials. This model helps one explain the relationships between surface stratigraphy, geomorphology, hydrology and soils. However, elements of Hack's and Conacher and Dalrymple's models can make important contributions in many cases.

All soils in a rolling landscape do not have the same history, and probably the most important part of the Ruhe model is that all soils on the landscape are not the same age. Ruhe's landscape model easily combines with his simple landscape position classification (Ruhe, 1975). This combination allows one to predict accurately soil materials and relative moisture relations.

Ruhe's landscape model and landscape position are important in understanding process, and especially in helping understand soil variability on a local landscape.

The idea of a modal profile or normal profile dies hard. Yet if one accepts Ruhe's landscape model, all soils in the landscape and the hillslope systems are normal for that site. We need to change our ideas from a typifying pedon or modal profile to one of sequential change within the hillslope system.

APPLICATION

Several soil-geomorphic studies tested Ruhe's model in a wide range of climate and materials (Daniels & Gamble, 1967, 1978; Daniels & Jordan, 1966; Daniels et al., 1970, 1971; Gile, 1975a, 1975b; Gile & Hawley, 1966; Gile et al., 1981; Ruhe et al., 1967). Many other studies show that soil properties relate to site history (Craig et al., 1972; Knuteson et al., 1989; McCracken et al., 1989). The stable parts of the landscape have thick soils compared to those on the adjacent younger erosional surfaces. There is a strong, but not absolute, relationship between geomorphic surface and soils. Soils can cross geomorphic boundaries, especially in old Coastal Plain areas. Only by recognizing the influence of stratigraphy and geomorphology on soils can we understand the moisture and topographic relationships related to soil genesis. Without the above background, the genesis studies are little more than soil characterization of isolated profiles.

It is probable that no one model will work in all landscapes. Davis's ideas closely fit the relationships found in closed systems with only one surface (undissected till plains). However, in open systems with stream dissection and more than one surface, the models of Ruhe and of Conacher and Dalrymple appear to be most useful. Possibly parts of each model will help explain the relationships in many landscapes. For open systems, the senior author considers Ruhe's model the best for a wide range of conditions; however, this may be a bias developed over many years of association. As a predictive tool, Ruhe's model works well if combined with some features of Hack's and Conacher and Dalrymple's ideas.

Soils are part of the geomorphic expression of any site. However, one must exercise caution in using soil series or mapping units to interpret processes responsible for a specific slope. Soil properties are an expression of the site history. Soil map units or series are generalizations that occur on a wide range of sites with different geomorphic histories. Soil maps are not geomorphic maps; we definitely should not use soil orders as indicators of geomorphic history. We must guard against the circular argument that soil A occurs on surface B, and therefore all areas of soil A are on surface B. It may or may not be true, but most likely is not true.

REFERENCES

Conacher, A.J. and J.B. Dalrymple. (1977). *Geoderma*, 18:1–154.

Craig, R.M., R.J. McCracken, and R.B. Daniels. (1972). *Soil Sci.*, 114:486–492.

Dalrymple, J.B., R.S. Blong, and A. Conacher. (1968). *Annals of A. Geomorphol.*, 12:60–76.

Daniels, R.B. and E.E. Gamble. (1967). *Geoderma*, 1:117–124.

Daniels, R.B. and E.E. Gamble. (1978). *Geoderma*, 21:41–65.

Daniels, R.B., E.E. Gamble, and J.C. Cady. (1970). *Soil Sci. Soc. Amer. Proc.*, 34:648–653.

Daniels, R.B., E.E. Gamble, and J.G. Cady. (1971). *Adv. Agron.*, 23:51–88.

Daniels, R.B. and R.H. Jordan. (1966). *Physiographic History and the Soils, Entrenched Stream Systems, and Gullies, Harrison County, Iowa*. USDA Tech. Bull. 1348.

Davis, W.M. (1899). The geographical cycle. *Geog. J.*, 14:481–504. Republished in (1954). *Geographical Essays*. New York: Dover.

Forman, R.T.T. and M. Gordon. (1986). *Landscape Ecology*. New York: Wiley.

Gile, L.H. (1975a). *Soil Sci. Soc. Amer. Proc.*, 39:316–323.

Gile, L.H. (1975b). *Soil Sci. Soc. Amer. Proc.*, 39:324–330.

Gile, L.H and J.W. Hawley. (1966). *Soil Sci. Soc. Amer. Proc.*, 30:261–268.

Gile, L.H., J.W. Hawley, and R.B. Grossman. (1981). *Soils and Geomorphology in the Basin and Range area of Southern New Mexico—Guidebook to the Desert Project*. New Mexico Bureau of Mines and Mineral Resources, Memoir 39.

Hack, J.T. (1957). *Studies of Longitudinal Stream Profiles in Virginia and Maryland*. U.S. Geol. Survey Prof. Paper 294.

Hack, J.T. (1960). *Am. J. Sci.*, 258A:80–97.

Knuteson, J.A., J.L. Richardson, D.D. Patterson, and L. Prunty. (1989). *Soil Sci. Soc. Amer. J.*, 53:495–499.

McCracken, R.J., R.B. Daniels, and W.E. Fulcher. (1989). *Soil Sci. Soc. Amer.*, 53:1146–1152.

Meyer, D.L. and L.A. Kramer. (1969). *Agri. Eng.*, 50:522–523.

Moore, I.D. and G.J. Burch. (1986). *Soil Sci. Soc. Amer. J.*, 50:1294–1298.

Moore, I.D., G.J. Burch, and E.M. O'Loughlin. (1986). *Soil Sci. Soc. Am. J.*, 50:1374–1375.

Naveh, Z. and A.S. Lieberman. (1984). *Landscape Ecology*. New York: Springer-Verlag.

Penck, W. (1953). *Morphological Analysis of Land Forms*. London: MacMillan.

Rowe, J.S. (1984). *Understanding Forest Landscapes*. [Leslie L. Schaffer Lectureship in Forest Science] Vancouver, BC: Univ. of British Columbia.

Ruhe, R.V. (1975). *Geomorphology*. Boston: Houghton Mifflin.

Ruhe, R.V., R.B. Daniels, and J.G. Cady. (1967). *Landscape Evolution and Soil Formation in Southwestern Iowa*. USDA Tech. Bull. 1349.

Schumm, S.A., M.P. Mosley, and W.E. Weaver. (1987). *Experimental Fluvial Geomorphology*. New York: Wiley.

11 Rates of Denudation

METHODS OF MEASUREMENT

How fast does the landscape evolve? What are the erosion rates—are our hills everlasting and unchanging, or are they rapidly evolving? How can we apply ideas of landscape evolution to current problems of soil erosion? What will happen in the future if we do not control erosion? Geologists have estimated long-term erosion rates and lowering of the landscape, and their data can help us partially answer the above questions.

Geologists use several methods to estimate erosion rates. These include sediment volumes of dated rock units and the suspended load of rivers (Cleaves et al., 1970; Deither et al., 1988; Gilluly, 1964; Gilluly et al., 1970; Hack, 1979; Judson & Ritter, 1964; Matthews, 1975; Menard, 1961; Ruhe & Daniels, 1965; Warner, 1985). Young (1969) used four lines of evidence to estimate the rate of loss of material from the land surface:

1. Estimates of dissolved and suspended materials transported by rivers.
2. Sediments accumulated in reservoirs.
3. Measurement of surface processes on slopes including soil creep, surface-wash and landslides.
4. Radiocarbon dating and associated landforms that are later.

Much of the work has been in regions that include several kinds of bedrock and climates. Considerable data are now available from both large and small areas. These data allow us to evaluate the rate of chemical weathering and physical erosion of particulate matter. We will limit our discussion of rates of denudation to investigations of sediment movement, redistribution and loss within landscapes.

EROSION RATES

Data of Young (1969) and Schumm (1963) give us some idea of how erosion rates apply to our ideas of soil development. Table 11.1 shows the erosion rates from the analyses of Young and Schumm for normal and steep topography.

By applying these data to the North Carolina Piedmont and Mountains it is possible to illustrate the amount of landscape lowering (Table 11.2). From

TABLE 11.1. Estimated Erosion Rates for Normal and Steep Topography

	Normal	Steep
Young (1969)	46 mm/1,000 yr	500 mm/1,000 yr
Schumm (1963)	72 mm/1,000 yr	915 mm/1,000 yr
Young (1969)	22,000 yr/meter	2,000 yr/meter
Schumm (1963)	14,000 yr/meter	1,100 yr/meter

TABLE 11.2. Estimated Lowering of the North Carolina Piedmont and Mountains During the Last 2 Million Years

	Piedmont	Mountains
	Landscape Lowering	
Young (1969)	92m (302ft)	1,000m (3,300ft)
Schumm (1963)	143m (470ft)	1,818m (6,000ft)
Hack (1979)	79m (260ft)	
	Soil Loss (Tons/Acre/Year)	
Young (1969)	0.5	5.3
Schumm (1963)	0.7	9.6
Hack (1979)	0.4	n/a

Table 11.2, it is also possible to relate geologic erosion to the allowed annual loss from cultivated soils.

Caution in Interpretation

The above erosion rates cited are subject to considerable error. These data show there has been much lowering of the landscape during the last two million years. The rates in mountainous areas are similar to what we consider to be allowable rates for agricultural erosion. There are several difficulties in applying the rates to the present landscapes.

1. A straight line extrapolation in time does not allow for the unsteady rates of erosion (episodic). There is a greater interval between deposits than the time required for deposition. Erosion and the accompanying sedimentation are not uniform processes; they go "steady by jerks."
2. In 225 million years, erosion has removed or reworked older sediments. The total volume of sediment may represent only a small part of that deposited during the interval involved. The erosion rates estimated by sediment volumes probably are somewhat less than the actual rate.

3. Underestimation of traction or bedloads is likely because they are difficult to measure.
4. Dissolved load does not always mean an equal lowering of the land surface. Isovolumetric weathering, especially in bedrock areas, would reduce the lowering of the land surface.

Annual erosion rates estimated by sediment volumes or other criteria are lower than estimates of erosion from cultivation. All these estimates suggest the entire landscape lost 0.5 to 8.8 m of material during the last 10,000 years. The mean rate from Hack's (1979) work is 3.7 m in 10,000 years. This suggests that no 1 m thick pre-Holocene soil exists today, even at half the above rate. Other evidence (Ruhe et al., 1967) shows that parts of the landscape are stable, and pre-Holocene soils exist on flats. Hack's (1979) data suggest that all soils on sloping landscapes are less than 5,000 years old, if soil removal is uniform.

The above interpretations may not be too far from fact. Erosion is not always uniform across the landscape if Ruhe's and others' ideas of landscape evolution are correct. This suggests that local erosion rates can be two to three times higher than the average. The gently sloping to level broad interfluves may have remained unchanged in elevation, although all have had some material removed by solution.

LATE HOLOCENE EROSION RATES

A difficulty with projecting erosion rates is that the record becomes less distinct with increasing time. The record is incomplete and few details are clear. The younger events are complete, or the missing parts are in similar deposits within the study area. A study of Holocene erosion rates allows us to approximate the processes responsible for the event.

General Relationships

Erosion rates vary with groundcover, climate, relief, rock-type, and location of measurement within the stream system. Very generalized data from large areas establish useful relationships for predicting erosion rates. Langbein et al. (1949) show that mean annual run-off decreases with increasing temperature (Fig. 11.1).

Erosion decreases exponentially with increased vegetative cover under the same conditions (Fig. 11.2). Small erosion plot data confirm this relationship. Cover is the major tool conservationists use to reduce erosion to acceptable levels.

Langbein and Schumm (1958) found that sediment yield increased to a maximum from 12 to 15 inches of *effective* annual precipitation. Effective precipitation is the amount required to produce a given run-off at 50 degrees Fahrenheit. Sediment yield decreased above 15 inches precipitation before

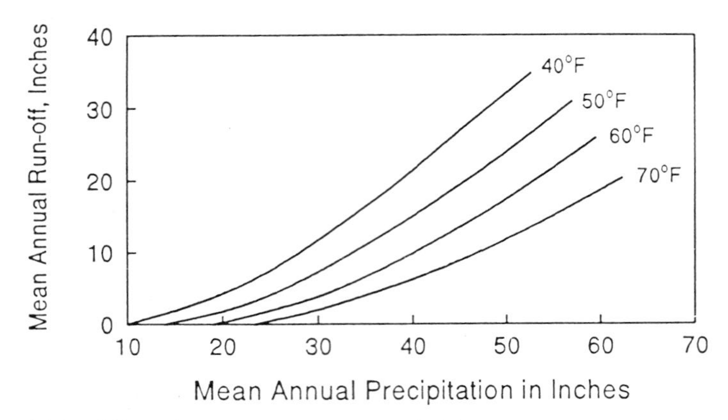

FIGURE 11.1. *Effect of temperature on run-off.* After Langbein et al., 1949, Fig. 2, p. 8.

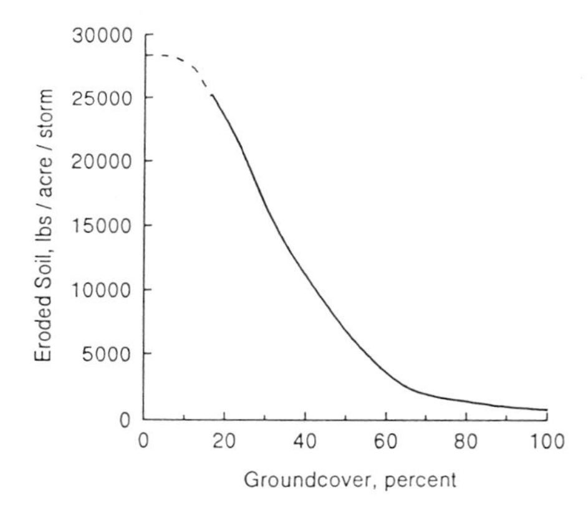

FIGURE 11.2. *Erosion vs. groundcover.* From Noble, 1965, Fig. 4, p. 118.

leveling off at 35 to 40 inches (Fig. 11.3). The large watersheds with moderate to low relief limit general application of the above data. Fournier (1960) shows two peaks of sediment yield in a similar study, but he did not include arid areas. The data shown in Figure 11.3 are a reasonable approximation of sediment yield in large watersheds of low or moderate relief.

Application to high-relief areas may have considerable error because watershed relief influences erosion rates. Schumm (1977) used watershed relief/length ratio as a measure of potential sediment yield (Fig. 11.4). These data show that sediment yields increase with higher relief per unit channel length. Low-relief areas with large proportional channel lengths have low sediment yields. Erosion increases exponentially with watershed relief/length ratio. The major inflection occurs at a ratio of 0.05 (Fig. 11.5).

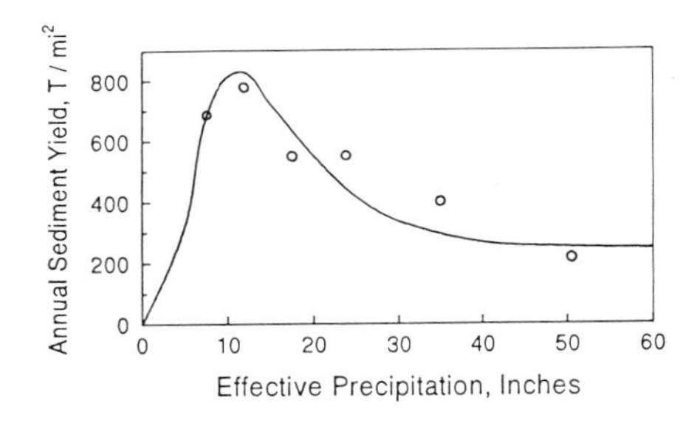

FIGURE 11.3. *Sediment yield and climate.* After Langbein and Schumm, 1958, Fig. 2, p. 1077. Republished by permission from American Geophysical Union.

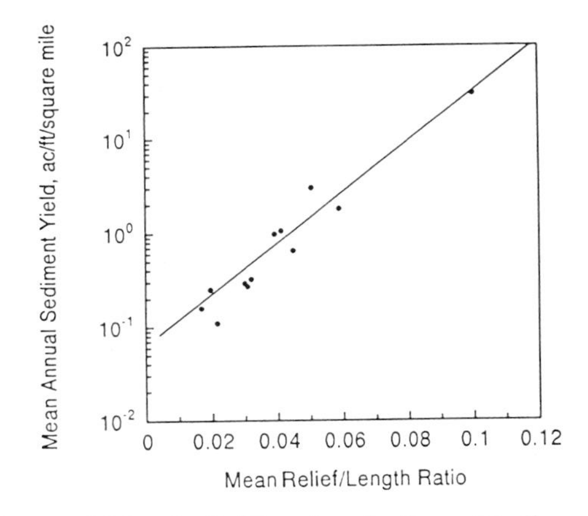

FIGURE 11.4. *Sediment yield and relief/length ratio.* From Hadley and Schumm, 1961, Fig. 30, p. 173.

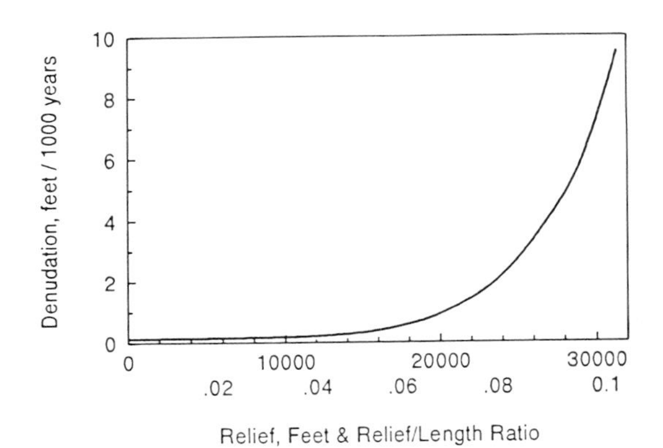

FIGURE 11.5. *Relationship between denudation rates to relief/length ratio and drainage-basin relief.* Schumm, 1963, Fig. 2, p. H6.

Watershed drainage density and lithology control sediment yield (Tables 11.3 and 11.4). Soil permeability and regolith cover control drainage density. Drainage density is high in resistant rocks and slowly permeable soils.

Erosion and sediment yields increase with *decreasing* size of drainage basin (Fig. 11.6; Table 11.4), and this increase or decrease with basin size is important in extrapolating field observations and other data. Exit-point data from large drainage basins probably have a tenuous relationship with the major erosive part of the watershed.

Each watershed has erosional and depositional elements. The valley slopes and rounded interfluves (Ruhe's model) are erosional areas. The depositional element includes the lower valley slopes and floor. Sediment measured at the mouth of the watershed of a nontrenching stream must ultimately come from the erosional element. Some sediment may come from the banks and bed of the channel system within the dominantly depositional element. One should locate valley slope erosion studies high in the watershed, not near the mouth.

Small Watershed Studies

Alluvial fills are nature's garbage dumps. These fills retain material useful in unraveling the erosional and depositional history of the adjacent area. This is especially true in the small (a few hectares) watersheds. For best results, the study must include connecting fills in a larger watershed to help understand the areal relations. Also, the larger watersheds there may retain a better selection of datable material.

TABLE 11.3. Sediment Yield vs. Lithology

Stratigraphic Units (formation)	Mean Infiltration (in./hr)	Sediment Yield (ac/ft/sq. mile)	Drainage Density (miles/sq. mile)
Wasatch	9.2	0.13	5.4
Lance	5.0	0.5	7.1
Fort Union	1.3	1.3	11.4
Pierre Shale	1.0	1.4	16.1
White River Group	0.18	1.8	258

From Hadley & Schumm, 1961, Table 2, p. 144 and p. 175.

Ruhe's study (Ruhe et al., 1967, pp. 149–155) of a first-order stream valley is an example of the detail needed for small watershed investigations. The area's sediments are clay loam till capped by Wisconsinan loess. Loess occupies the sloping spur ridges. The valley slopes are till with slopes of 8 to 12 percent grading to the alluvial fill in the ephemeral channel. A stone line marks the base of the alluvial fill. Figure 11.7A gives the longitudinal section of the channel and

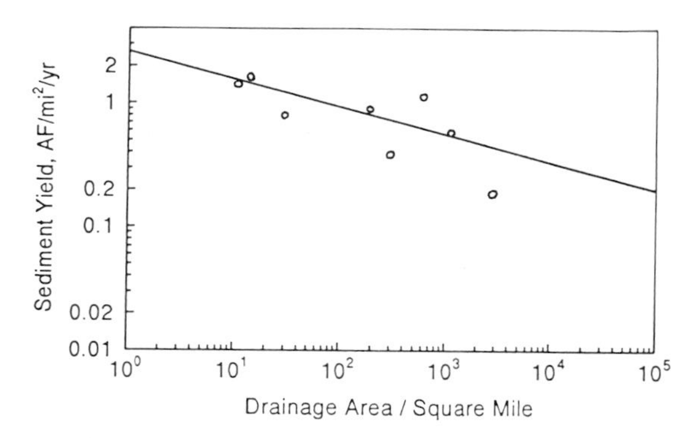

FIGURE 11.6. *Sediment yield vs. drainage area.* From Strand, 1975, Fig. 5, p. 13.

TABLE 11.4. Sediment Yield as Influenced by Drainage Density

Drainage Area (sq. mile)	Relief/ Length	Drainage Density	Sediment Yield (ac. ft/sq. mile)	Mean Annual Run-off (ac. ft/sq. mile)
1.18	0.023	0.6	0.6	16.3
0.87	0.026	2.1	1.7	22.1
0.33	0.043	3.1	3.7	44.5
0.29	0.046	5.6	6.8	63.4
1.04	0.035	1.3	0.9	20.5
0.25	0.036	3.0	2.2	36.8

Modified from Schumm, 1969, Table 1, p. 61, by permission from American Society of Agricultural Engineers.

its alluvial fill. Figure 11.7B is two cross-sections of the fill and pre-fill surface. The assumed age of the basal fill is 6,800 years. The upper 22 inches of the fill is post-settlement alluvium that is <125 years. From the cross-sections and contour maps the following data are available:

Source area before filling	191,754 ft^2 or 4.4 acres
Source area after filling	75,263 ft^2 or 1.73 acres
Pre-settlement alluvium	22,171 yds^3 or 3.3 yds^3/yr
Post-settlement alluvium	1,394 yds^3 or 11.2 yds^3/yr

Erosion rate pre-settlement alluvium for the 3.05 acre watershed equals 0.53 tons/acre/year. Pre-settlement erosion removed 3.1 feet of material from the valley slopes and deposited it in the adjacent fill. Post-settlement erosion removed 0.5 foot of material from the valley slopes.

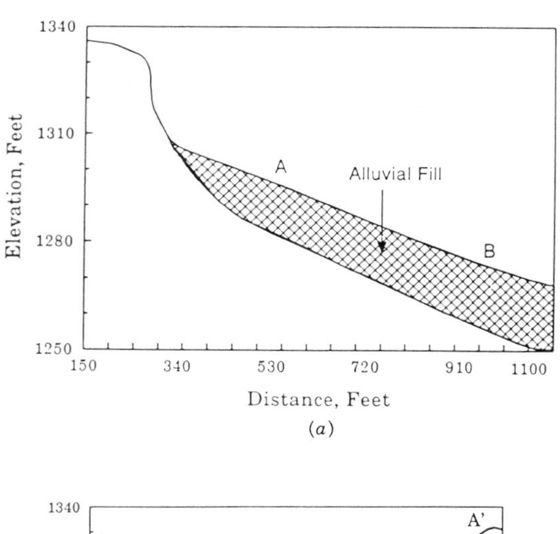

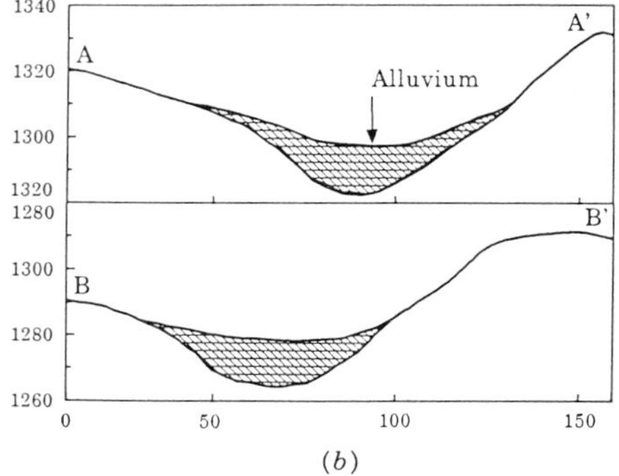

FIGURE 11.7. *Side valley first-order channel.* A. After dissection. B. After filling. From Ruhe et al., 1967, Fig. 61, p. 161.

Daniels and Jordan (1967) measured the volume of sediments from a 35.2 acre watershed in Harrison County, Iowa. The watershed has 7.0 acre of Mullenix fill exposed at the surface. The 23.2 acre valley slopes grade to the top of the Mullenix fill. About 5 acres are uneroded interfluve. The Mullenix alluvium in the watershed has 4,029,000 cubic feet, or 92.49 acre feet. The volume of the Mullenix fill requires an erosion depth of 4.0 feet on the valley slopes. This erosion and deposition of the Mullenix fill was between 250 and 1,800 years ago.

If we assume uniform erosion for 1,500 years, the loss was 0.23 foot in 100 years. Deposition of 79% of the fill at the Mullenix-type section occurred during the last 1,100 years. Using these data, erosion removed about 0.36 foot from the valley slopes every 100 years from 250 and 1,100 years ago. We do not know the sediment delivery ratio, so the rates computed are a minimum in this open system. The estimated removal of 4.0 feet from the entire erosion surface may be 20 to 50 percent low. The annual erosion rates in tons/acre/year are shown

TABLE 11.5. Annual Erosion Rates

Years	Removal	Rates
0–115	0.19 foot	= 0.02 in./yr = 3 T/ac/yr (post-settlement)
250–1,100	3.15 feet	= 0.04 in./yr = 6 T/ac/yr
250–1,800	3.98 feet	= 0.03 in./yr = 4.5 T/ac/yr

in Table 11.5. Erosion rates between 250 and 1,800 years ago were under virgin grassland conditions with little disturbance from Amerindian cultures. These data clearly illustrate the uneven erosion rates over time. These data illustrate that geologic erosion in agricultural areas equals or exceeds allowable rates for maintenance of soil productivity. This calls into question the validity of our "allowable rates" of erosion.

REFERENCES

Cleaves, E.T., A.E. Godfrey, and O.P. Becker. (1970). *Geol. Soc. Amer. Bull.*, 81:3015–3032.

Daniels, R.B. and R.H. Jordan. (1966). *Physiographic History and the Soils, Entrenched Stream Systems, and Gullies, Harrison County, Iowa*. U.S. Dept. Agri. Tech. Bull. 1348.

Deither, D.P., C.D. Harrington, and M.J. Aldrich. (1988). *Geol. Soc. Amer. Bull.*, 100:928–937.

Fournier, F. (1960). *Climat et erosion: La relation entre l'erosion du sol par l'eau et les precipitation atmospheriques*. Paris: Presses Universitaire de France.

Gilluly, J. (1964). *Geol. Soc. Amer. Bull.*, 75:483–492.

Gilluly J., J.C. Reed, and W.M. Cady. (1970). *Geol. Soc. Am. Bull.*, 81:353–376.

Hack, J.T. (1979). *Rock Control and Tectonism—Their Importance in Shaping the Appalachian Highlands*. U.S. Geol. Survey Prof. Paper 1126B.

Hadley, R.F. and S.A. Schumm. (1961). *Sediment Sources and Drainage Basin Characteristics in Upper Cheyenne River Basin*. U.S. Geol. Surv. Water Supply Paper 1531-B:137–196.

Judson, S. and D.F. Ritter. (1964). *J. Geophys. Res.*, 69:3395–3402.

Langbein, W.B. and S.A. Schumm. (1958). *Amer. Geophys. Union Trans.*, 39:1076–1084.

Langbein, W.B., et al. (1949). *Annual Runoff in the United States*. U.S. Geol. Survey Circ. 52.

Matthews. W.H. (1975). *Am. J. Sci.*, 175:818–824.

Menard, W.H. (1961). *J. Geol.*, 69:1255–161.

Noble, E.L. (1965). *USDA Misc. Publ.*, 970:114–123.

Ruhe, R.V. and R.B. Daniels. (1965). *J. Soil and Water Conserv.*, 20:52–57.

Ruhe, R.V., R.B. Daniels, and J.G. Cady. (1967). *Landscape Evolution and Soil Formation in Southwestern Iowa*. U.S. Dept. Agri. Tech Bull. 1349.

Schumm, S.A. (1963). *Disparity Between Present Rates of Denudation and Orogeny*. U.S. Geol. Survey Prof. Paper 454H:H1–H13.

Schumm, S.A. (1969). *Amer. Soc. Agri. Eng. Trans.*, 12:60–68.

Schumm, S.A. (1977). *The Fluvial System*. New York: Wiley.

Strand, R.I. (1975). Bureau of Reclamation procedures for predicting sediment yield. *Agricultural Research Service*. ARS-S-40:10–15.

Warner, R. (1985). Themes in Australian fluvial geomorphology. In *Geomorphology: Themes and Trends* (pp. 85–101). Ed. by A. Pitty. Totowa, NJ: Barnes & Noble.

Young, A. (1969). *Nature*, 224:851–852.

12 Streams

INTRODUCTION

Streams are effective agents of change in soil morphology because soils are dynamic systems that continually evolve in response to their environment. Streams effect soil systems by their influence on soil drainage. Soils in poorly drained sites will have gray B horizon colors (10YR6/2–7/2). These gray colors are from reflected light of iron-poor sand, silt and clay particles, and not from ferrous iron (Daniels et al., 1961). Reduced soils with free iron have matrix colors on the Munsell gley chart (Daniels et al., 1961, 1973). Stream dissection can change the poorly and very poorly drained sites to well-drained soils by lowering the water table.

Reduced iron is very soluble and leaves the sediment and soils in drainage water. The reduced systems that have some iron before stream dissection may change considerably after dissection moves water through the material. The soils and sediments may turn gray with continued iron removal under reducing conditions and slow drainage. The same soils become yellowish brown to gray under oxidizing environments with rapid drainage. The areas with gray soils before dissection will remain gray because they have little free ferric iron to produce brown or red colors. These well-drained soils have relict gley and retain the morphology of poorly drained sites. Stream dissection can change the oxidation-reduction and leaching regimes of a landscape that result in local differences in soil iron distribution (Daniels et al., 1975).

In the above example, the dissecting stream alters the drainage of the soil landscape and, in effect, "drives" the system genesis. For example, a broad poorly drained flat under a given climatic regime will remain wet if streams do not dissect it. Examples of incipient dissection from artificial drainage are the Atlantic Coastal Plain and the Red River Valley in North Dakota and Minnesota. The interval since drainage has been too short to affect soil morphology, but the soil processes have changed.

Water must move through the sediment and soil for leaching and deep soil formation to develop. The dynamics of the hydrologic and soil systems change once erosion surfaces from the streams start "nibbling at the edges" of the interstream flat. Dissection increases water movement through the sediment, and deep leaching and soil formation are then possible. Climatic shifts or other major changes are not necessary to modify the soil-sediment dynamics. Stream incision is enough.

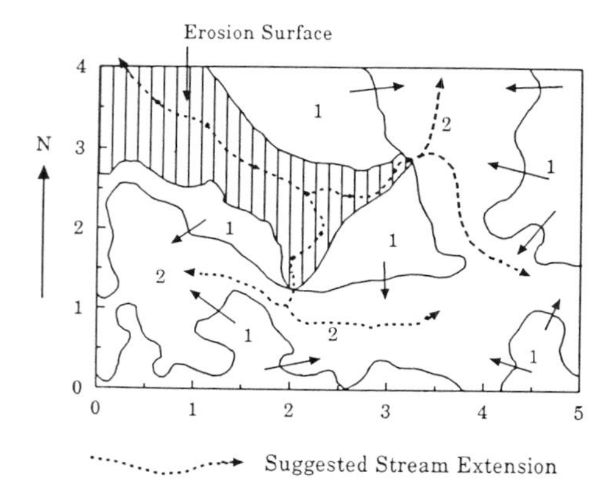

FIGURE 12.1. *Potential area of channel extension* (dashed lines) *into a very gently undulating coastal plain constructional surface* (plain or slanting haucher). Hauchered area is present dissected area. Numeral 1 designates areas with well to somewhat poorly drained soils. Poorly and very poorly drained soils, wetlands, are indicated by the numeral 2. The areas within 1 are slightly higher than the adjacent low wet area within 2. Modified from Karnowski et al., 1974, sheet 48.

The stream system controls dissection of the landscape. A degrading stream, for whatever reason, extends headward into undrained, nearly level areas along the water-collecting lows. For example, dissection of nearly level coastal plain areas can occur during run-off from hurricanes. Water collects in the low areas and eventually flows to a channel head.

The following cartoon (Fig. 12.1) is an idealized version of the undissected areas of the Pitt County, North Carolina Coastal Plain near Frog Level. Wet soils (Aquults) occupy the lows and depressions of the very gently undulating upland. Local relief is less than a meter in many areas. The present stream channels always head in or near a large continuous area of wet soil. No visible channel exists within the wet soil delineation.

During high-intensity rains, run-off from the higher moderately well-drained soils collects in the lower poorly drained soils. During hurricanes, several centimeters of water can collect in the lows. When the ponded water becomes deep enough, it flows from one low to another and eventually to the channel head. Dissection from surface flow starts at the outlet of the low areas. Headward movement of the head cut follows the source of moving water or where water collects, as indicated by the dashed lines in Figure 12.1.

Erosion of the low wet areas is the major reason few remnants of poorly drained soils persist in dissected coastal plains. On an undissected surface, the well-drained soils are micro-ridges, and the wetter soils the footslope or micro-valleys. On the dissected surface, the well-drained micro-ridges become the interfluves. Erosion removes the former poorly drained soil, and that area becomes a valley slope or bottom.

DEGRADING AND AGGRADING SYSTEMS

The stream system is a very important geomorphic unit because it controls the local slope from the interfluves to the channel. In a degrading or trenching stream system, the adjacent valley slopes and tributary channels have an increasing erosion potential. Channels and valley slopes that graded to a higher base-level now are subject to attack from the new erosion surface. The initial adjustment of the valley slopes and tributaries will be from the main stream headward. The new surface rising from the trenched channel will not be the same age throughout a large watershed.

A degrading system may require considerable time to move throughout the watershed. For example, the basal part of the alluvial fill in one stream system of northwestern Mississippi has a range of 2 to 3 thousand years (Grissinger & Murphey, 1984; Grissinger et al., 1982). The age of alluvium decreases with increasing distance of the site from the controlling river. Thus, incision and subsequent filling are not contemporaneous over the entire watershed. This does not mean that the trenching must always start at the mouth of the watershed; it can start any place in the system. The discontinuous gullies of the west (Schumm, 1977) are examples of trenching starting anywhere in the system.

An aggrading system buries the lower valley slopes in alluvium. This shortens the bounding slopes that grade to the top of the adjacent fill. In the rising base-level, the alluvial fill moves toward the source, eventually mantling concave valley slopes. Again, the alluviation does not have to start at the mouth of the watershed, but the mouth is the most likely starting point for deposition.

In trenching systems, the potential is for considerable erosion of the alluvial fill and the adjacent upland (Fig. 12.2A). In aggrading systems most of the erosion is on the valley slopes. A rising base-level reduces relief and results in less headward extension of the bounding erosion surfaces (Fig 12.2B). In

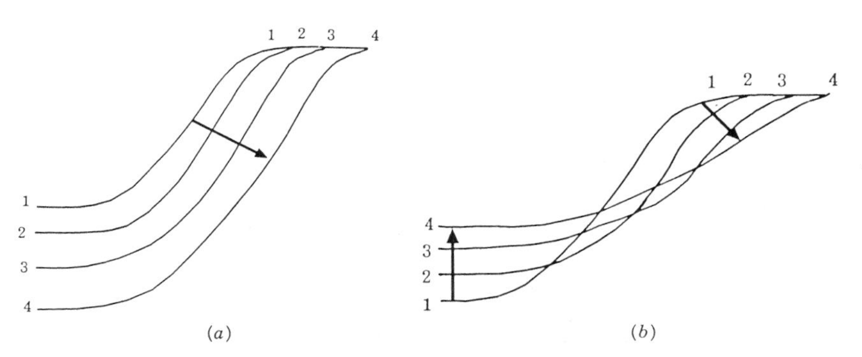

FIGURE 12.2. *Suggested major erosional and depositional areas adjacent to stream channels* in (A) Degrading channels, and (B) Aggrading channels.

Figure 12.2A the channel will remove material delivered to it, and in Figure 12.2B much of the material remains within the watershed.

Examples

In both degrading and aggrading channel systems, the uplands erode and new surfaces develop. Aggrading systems may retain some parts of the older surfaces because upland slope recession is less than in a deeply degrading unit. Data from the Deforest Formation in southwest Iowa illustrates the possible sequence of fills from an aggrading and degrading channel (Daniels & Jordan, 1966).

The Deforest Formation occurs where about 21 m of late Wisconsinan loess mantles a mid to early Pleistocene till. The loess mantles a dissected landscape, and the present river valleys occupy the same position as the pre-loess valleys. Trenching of the valley occurred sometime after loess deposition (Fig. 12.3) and was deep enough to expose unoxidized and unleached till.

Deposition of the Deforest Formation was during the next 11,000 years. From about 11,000 to 1,800 years bp, the stream channels were aggrading. After 1,800 years bp the channels alternately degraded and filled, but never to the earlier level (Bettis, 1990). Today all channels have trenched in response to deepening of the Willow River (Daniels, 1960). Within a watershed, the valley slopes grade to the top of older fills, not the modern channels.

Throughout the Thompson Creek watershed, bed number 3 forms a nearly continuous terrace on the south-side of the valley (Fig. 12.4). The north-facing valley slopes grade to the top of bed 3 in much of the watershed. The opposing south-facing slopes grade to the top of bed 4. Thus the north-facing slopes are 1,800 years old, and the opposing south-facing slopes are about 500 years old.

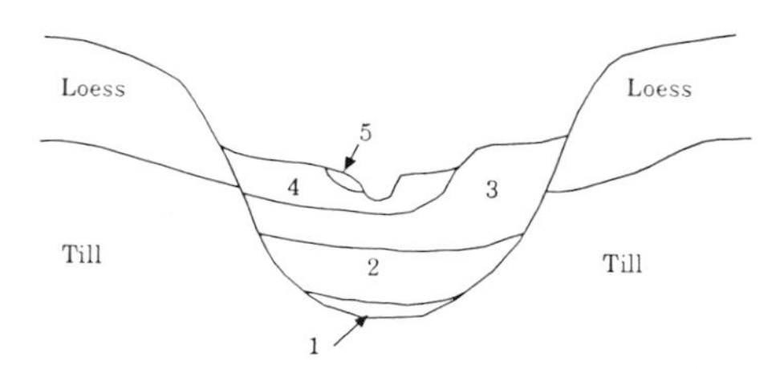

FIGURE 12.3. *Idealized cross-sections of the De Forest Formation* in the Thompson Creek Valley, Harrison County, Iowa. Modified from Daniels and Jordan, 1967, Fig. 14, p. 25, and personal experience. 1 = Soetmelk member of the DeForest Formation as defined by Daniels & Jordan, 1966 (see Bettis, 1990). 2 = Watkins bed. 3 = Hatcher bed. 4 = Mullenix bed. 5 = Turton bed as redefined by Bettis, 1990.

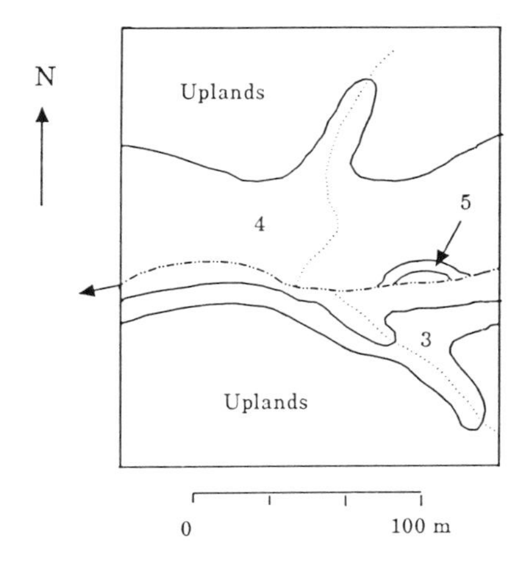

FIGURE 12.4. *Idealized areal distribution of the Mullenix and Hatcher fills at the surface in a Thompson Creek.* Modified from Daniels and Jordan, 1967, Fig. 15, p. 25, and personal experience. 3 = Hatcher bed. 4 = Mullenix bed. 5 = Turton bed.

Truncated Ephemeral Channels

The southern Piedmont is an example of a highly variable, old erosional landscape. Rounded interfluves occur on major divides, and steep valley slopes occur next to some stream channels. A first impression is that the entire landscape grades to the adjacent stream system. The Piedmont appears as an ideal landscape to apply the ideas of Hack (1960). All parts of the landscape are sloping and may be in equilibrium with each other. This infers that all parts of the landscape are about the same age. A close examination of ephemeral channels, however, may refute the idea that the Piedmont surfaces are the same age. The thalwegs of first- and second-order ephemeral channels may grade to a level above the present controlling stream.

Hack (1957) studied stream profiles in Virginia and Maryland and developed methods to quantify the downstream changes in channel slope. An idealized graded channel would have a constantly decreasing slope downstream. The stream profile plots as a straight line on semilog paper if particle size remains constant. The stream profile is a straight line on double-log paper where the particle size changes systematically downstream (Hack, 1957). Subtle changes in channel slope are easy to locate on log-paper plots. This method of locating slope changes is more precise than curved arithmetic plots. A log plot quickly allows one to find potentially critical areas in the channel.

Parts of the Puerto Rico Highlands (Daniels et al., in press), southern Brazil (Lepsch et al., 1977) and the North Carolina Piedmont (Schonenberger et al., 1985) have first-order and second-order channels with convex reaches. The various reaches of channel or alluvial fills plot as straight lines. The channel below the convexity is steeper than the one above (Fig. 12.5A and B). Valley slopes grade smoothly to the channel above the convexity. Downstream from

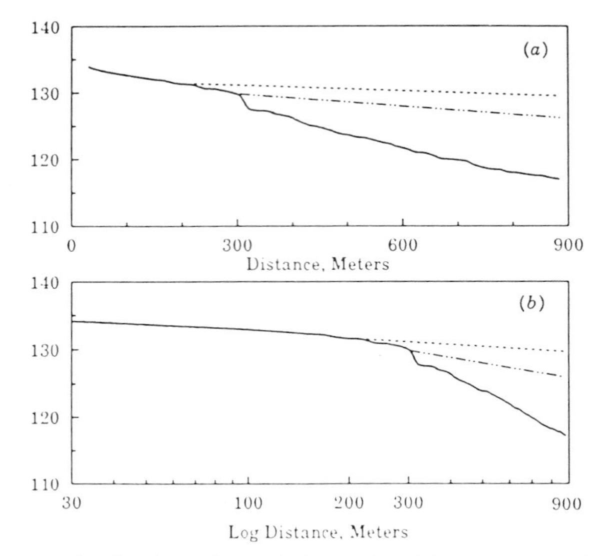

FIGURE 12.5. Longitudinal profiles of channels with convex reaches plotted (A) Arithmetic scale and (B) Log scale.

the convexity, the steeper valley slopes grade to the lower channel (Fig. 12.6A and B). Slopes grading to a channel are part of that geomorphic surface.

A convex shoulder marks the slope change from the higher to the lower surfaces, or the surfaces related to each reach. One can trace these shoulders for some distance within, and frequently throughout, the watershed (Fig. 12.6A). We believe the valley slopes and channels to which they grade are one geomorphic surface. Our interpretation suggests that the Piedmont has several surfaces, each separated by a convex shoulder (Schonenberger et al., 1985). These interpretations require that bedrock outcrops or similar features do not control the location of the convex reach. If bedrock controls the location of the convex reach, then all surfaces are the same age.

To estimate stream incision, extrapolate the channel slope to the valley center (Fig. 12.5B) and measure its altitude. This method assumes that a channel in pseudoequilibrium would plot as a straight line. If the channel is a relict graded to a higher base-level, then the adjacent slopes are a relict landscape. The boundary between different landscapes is the shoulder or change in slope (Fig. 12.6A and B).

INTERPRETATION FOR SOILS

Morphological expression of soil horizons is very dependent upon soil moisture regime. Mineral material changes little, and an organic soil may develop if little water moves through the soil and sediments on a broad flat. The same

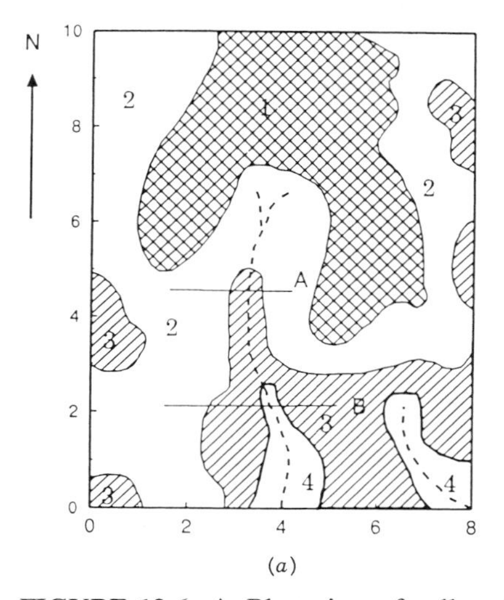

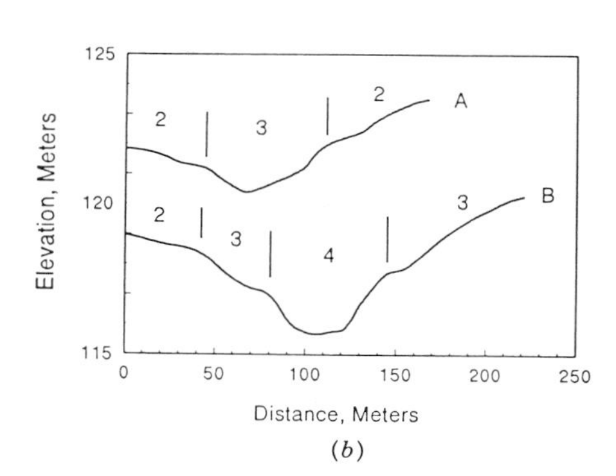

FIGURE 12.6. A. Plan view of valley slopes and channel with convexity. B. Cross-section of valley slopes above and below the convexity.

or similar material with excellent underdrainage may develop a soil with well-expressed mineral horizons. Weathering can change the mineralogical and chemical composition of the soil parent material.

Examples of these differences are the organic soils in Hofmann Forest (Daniels et al., 1977), and the Greenville soils (Rhodic Paleudults) in southeastern Alabama on similar materials (Mattox, 1975). The organic soils of Hofmann Forest occur on broad interstream divides with a confining bed within 2 or 3 m of the surface. Very little water moves into the underlying mineral materials (Daniels et al., 1978), and the materials have gley colors (5GY4/1 or equivalent). The Greenville soils in Escambia County, Alabama, are on a dissected plateau and are red to depths of 4 to 5 m. Water moves freely through these soils and the underlying sediments. Stream dissection and sediment hydraulic conductivity control the soil environments of Hofmann Forest and Escambia County, Alabama. In Hofmann Forest, a confining bed and a broad interstream divide make surface and subsurface water movement very slow. In the Greenville (Escambia County) environment, the wide interstream divide has little or no restriction to vertical water movement, at least where the Greenville soils form.

The stream system plays a major role in the processes leading to soil development wherever there is an excess of rainfall over evapotranspiration. The morphology of the soils reflects the current and past processes. There are many examples where the morphological expression of the soil is relict. Where this happens, one must be aware of the stream system and its possible influence.

Holocene erosion exposes gray loess on the valley slopes in southwestern Iowa (Fig. 9.2A). The gray colors developed under saturated conditions (Ruhe et al., 1955), but saturation does not occur today. Such conditions are not unusual in erosional landscapes. In North Carolina, typic Paleaquults can have a lower mean monthly water table than an aquic Paleudult (Daniels et al., 1987). The controlling factor is the soil's location in the hydrologic gradient as measured by its distance from the dissected edge of the Coastal Plain flat. Stream incision reduced the duration of saturation at the Paleaquult site. The poorly drained morphology remains because weathering has not produced enough free iron to color the B horizon. Similar soil differences occur in well-drained soils (Table 12.1). Both soils are within the range of characteristics of the Norfolk Series. The pedon in the divide center has 10YR7/3 mottles in the upper Bt. A pedon next to the edge of the upland flat has a strong brown Bt horizon. The differences in water-table levels during the wet months are large.

Reliance upon subsurface soil colors alone to classify soil drainage conditions can result in misclassification. One must consider the location of the soil within the landscape, subsurface hydrology, management practices, site history and seasonal distribution of precipitation to classify soil drainage correctly. For example, removal of forest cover from nearly level soils underlain by impervious sediments can raise and prolong the duration of seasonal water tables. The combined effects of reduced canopy interception and reduced

TABLE 12.1. Mean Monthly Water Tables of Typic Paleudults in Relation to Location from Edge of Interstream Flat

Location	Divide Center	Edge of Upland Flat
B Horizon Morphology	*10YR7/3 Mottles 38–66 cm*	*7.5YR Bt*
Month	Depth to Water Table, Centimeters	
January	28	102
February	24	86
March	47	126
April	62	138
May	70	158
June	96	159
July	112	151
August	134	151
September	235	150
October	84	204
November	183	179
December	356	81

From Daniels et al., 1987, sites 15 and 18, Table 3, p. 24.

evapotranspiration allow more water to enter and remain in the soil. In some cases, reestablishment of the native forest vegetation becomes very difficult. Ridging the soil is necessary to aid seedling establishment. Such mechanical site preparation is a common practice for loblolly pine in the Atlantic and Gulf Coastal Plains.

The junior author mapped soils derived from talc in Skagit County, Washington (Thornton silt loam, fine-silty, serpentinitic, nonacid, mesic Aquic Xerorthents). The matrix color of the parent material was 10YR8/1 within 2 inches of the soil surface. The white color is a relict from the parent material. Soil water was never within 3 feet of the surface during a year's monitoring under a forest canopy. The correlators classified the soil as being "somewhat poorly drained." Soil color was the only criterion for the decision. When the native Sitka spruce forest was clear cut, water ponded on the soil surface in winter. Adjacent soils of the same series, but with a forest cover, did not have ponded surface water. One must understand the system dynamics and the effects of temporary disturbances when classifying or interpreting soil morphology.

Stream dissection partly controls the distribution of organic soils in the North Carolina Coastal Plain. In Hofmann Forest the organic materials thin and disappear as one approaches the head of a drainage channel (Daniels et al., 1977). The organic soils occur away from channels where a confining bed underlies the surface sediments. The aquitard and distance from an outlet make drainage of the area very slow. In these positions, rainfall alone sustains the lakes (Daniels et al., 1977). Obviously, one must consider the influence of the local stream system on the dynamics of the soil system.

REFERENCES

Bettis, A.E., III. (1990). *Holocene Alluvial Stratigraphy and Selected Aspects of the Quaternary History of Western Iowa*. Midwest Friends of the Pleistocene 37th Field Conference Guidebook: 1–16.

Daniels, R.B. (1960). *Am. J. Sci.*, 258:161–176.

Daniels, R.B., F.H. Beinroth, L.H. Rivera, and R.B. Grossman. (in press). *Geomorphology and Soils in an Area of East-Central Puerto Rico*. Soil Survey Invest. Bull.

Daniels, R.B., E.E. Gamble, and S.W. Buol. (1973). Oxygen content in the ground water of some North Carolina Aquults and Udults. In *Field Soil Moisture Regimes* (pp. 153–166). Ed. by R.R. Bruce, K.W. Flach and H.M. Taylor. Soil Sci. Soc. Amer. Special Pub. 5.

Daniels, R.B., E.E. Gamble, S.W. Buol, and H.H. Bailey. (1975). *Soil Sci. Soc. Amer. Proc.*, 39:335–340.

Daniels, R.B., E.E. Gamble, L.A. Nelson, and A. Weaver. (1987). *Water-Table Levels in Some North Carolina Soils*. Soil Survey Invest. Rpt. 40, USDA Soil Conservation Service.

Daniels, R.B., E.E. Gamble, W.H. Wheeler, and C.S. Holzhey. (1977). *Soil Sci. Soc. Amer. J.*, 41:1175–1180.

Daniels, R.B. and R.H. Jordan. (1966). *Physiographic History and the Soils, Entrenched Stream Systems, and Gullies, Harrison County, Iowa*. USDA Tech. Bull. 1348.

Daniels, R.B., G.H. Simonson, and R.L. Handy. (1961). *Soil Sci.*, 91:378–382.

Daniels et al. (1978). *Water Movement in Surficial Coastal Plain Sediments, Inferred from Sediment Morphology*. North Carolina Agri. Exp. Sta. Tech. Bull. 243.

Grissinger, E.H. and J.B. Murphey. (1984). *J. Miss. Acad. Sci.*, 24:89–96.

Grissinger, E.H., J.B. Murphy, and R.L. Frederking. (1982). Geomorphology of the Upper Peters Creek Catchment, Panola County, Mississippi. Part II: Within channel characteristics. In *Modeling Components of Hydrologic Cycle* (pp. 267–282). Littleton, CO: Water Resources.

Hack, J.T. (1957). *Studies of Longitudinal Stream Profiles in Virginia and Maryland*. U.S. Geol. Survey Prof. Paper 294.

Hack, J.T. (1960). *Am. J. Sci.*, 248A:80–97.

Karnowski, E.H., J.B. Newman, J. Dunn, and J.A. Meadows. (1974). *Soil Survey of Pitt County, North Carolina*. USDA Soil Conservation Service.

Lepsch, I.F., S.W. Buol, and R.B. Daniels. (1977). *Soil Sci. Soc. Amer. J.*, 41:104–109.

Mattox, M.G. (1975). *Soil Survey of Escambia County, Alabama*. USDA Soil Conservation Service.

Ruhe, R.V., R.C. Prill, and F.F. Riecken. (1955). *Soil Sci. Soc. Amer. Proc.*, 19:345–347.

Schonenberger, P.J., C.W. Smith, and T. Fox. (1985). Ephemeral channel roll overs (nick points) and their relationship to landscape development in the Piedmont of North Carolina. *Agron. Abs. ASA*: 197.

Schumm, S.A. (1977). *The Fluvial System*. New York: Wiley.

13 Hillslope Processes and Mass Movement

INTRODUCTION

Geomorphologists have actively studied slope development since the science started. The early approach to the problem developed theoretical or inductive models. Later approaches range from highly detailed quantitative studies of slope form to mathematical models based upon specific assumptions. The search continues. We now have much more sophisticated techniques and data, but it is doubtful if we are that much closer to understanding how a slope evolves than were the earlier scientists.

A common statement in the literature is that many slopes have changed very little in the last 10,000 to 20,000 years (Ciolkosz et al., 1989; Schafer & Hartshorn, 1965; Selby, 1982; Small & Clark, 1982). These authors recognize that erosion removed considerable material from some of these slopes. For soil genesis studies, it is important to know whether the slopes have been stable (essentially 0 erosion) or eroded. If the slopes erode, it is necessary to know when erosion started and, if it stopped, when. One should always question whether erosion ever stops on slopes.

It is important to geomorphologists to know that slope shape changes very little with erosion. Yet this information has little value to pedologists interested in soil genesis. Pedologists need to know when the slope stabilized because that is "time 0" for soil formation. If erosion and soil formation were equal, the pedologist and geomorphologist can use the same data. If the erosive periods were 6,000 to 2,000 years bp, then the data from the 10,000 year span is of little use in soil studies. For example, some geologists feel that only minor changes have occurred in the northeastern United States during Holocene (Schafer & Hartshorn, 1965). This idea leads Ciolkosz et al. (1989) to the conclusion that the age of these deposits is the age of the soils in the area.

Although some sediment or material weathering may date from the Pleistocene, it is doubtful if sloping landscapes have escaped erosion during the last 10,000 years. Denny and Goodlett (1956) estimated that tree throw plows the surface horizons once in 300 years. Most trees fall downslope and move large amounts of soil material a few centimeters to a meter. The exposed root ball allows rainfall and other forces to move material downslope. Much of the landsurface and associated soils in the glaciated northeast can be mid to late Holocene if data from Walker (1966) and Burras and Scholtes (1987) are applicable. Interpretation of soil and landscape data will differ considerably if one believes an area is uneroded or if it dates from mid to late Holocene.

HILLSLOPE EROSIONAL PROCESSES

Most scientists, at least those working in semiarid and humid regions, think of hillslope evolution by water erosion from overland flow. This is especially true of agricultural scientists because water erosion is of primary concern in cultivated areas. Forested areas with multi-layered canopies and thick litter layers have little evidence of overland flow, but removal of material is still possible (Table 13.1). In most nonforested areas some overland flow occurs during intense storms. Plant cover is a major, but not absolute, deterrent of erosion on slopes having some overland flow (Fig. 11.5).

Solution is a very effective but underrated process in removing material from a landscape. Solution does not always result in lowering of a surface, at least initially, because the rock and soil volume may remain constant (Cleaves et al., 1974). Solution and piping (tunnel erosion) are subsurface processes not requiring overland flow, and both require precipitation to exceed evapotranspiration. Channel erosion and the positive pressures developed at seeps require saturation and flow volumes large enough to detach and remove material.

Faunal activity and tree throw expose bare soil so raindrop splash and overland flow can sort and remove material. Creep is more effective on bare or nearly bare surfaces where roots do not increase soil strength. Alternate freezing and thawing effectively prepares soil material for movement on a bare surface.

TABLE 13.1. Processes Leading to Water Erosion on Permanently Vegetated Slopes with or without Overland Flow

Process	References
Solution	Cleaves et al., 1974; Pavich, 1986.
Piping	Finlayson & Statham, 1980; Selby, 1982.
Channel	Selby, 1982.
Seep spot	Zaslavsky & Sinai, 1981.
Faunal activity	Alvarado et al., 1981; Black & Montgomery, 1991; Bonnel et al., 1986; Hole, 1981; Hugie & Passey, 1963; Lyford, 1963; Thorp, 1949.
Tree throw	Denny & Goodlet, 1956; Denny & Lyford, 1963.
Creep	Selby, 1982.
Freezing & thawing	Small & Clark, 1982.

Rainfall

Raindrops are a primary cause of erosion if they can strike a bare surface. Raindrop diameter becomes larger with increasing rainfall, and the impact energy of raindrops increases with size and velocity (Fig, 13.1A and B). Raindrop impact breaks down aggregates and compacts the soil surface, which decreases infiltration and increases run-off (Finlayson & Statham, 1980).

Raindrops toss soil particles into the air both upslope and downslope. The downslope trajectories are longer than those upslope, so the mean particle movement is downhill. In level topography, the net erosion is zero from splash unless overland flow occurs. Raindrops can push soil particles about 50 mm by direct impact and can move larger fragments downslope by undermining them (Finlayson & Statham, 1980). Raindrops also produce turbulence in surface run-off, which produces more shear on the surface than laminar flow. Maximum erosion of bare surfaces occurs when run-off depth equals raindrop diameter, or between 3 to 6 mm.

Selby (1982) shows that the kinetic energy of raindrops is higher than that of the resultant run-off as overland flow. With a rain mass of R, and a terminal velocity of 8 m/s, the kinetic energy = 32R. If 1/4 of the rain is surface run-off at 1 m/s, its kinetic energy as run-off is R/8. In this example, the rainfall has 256 times more energy than the resultant run-off. Vegetative cover easily adsorbs the high kinetic energy of raindrops (Fig. 11.5), but any break in vegetative cover can result in erosion on slopes.

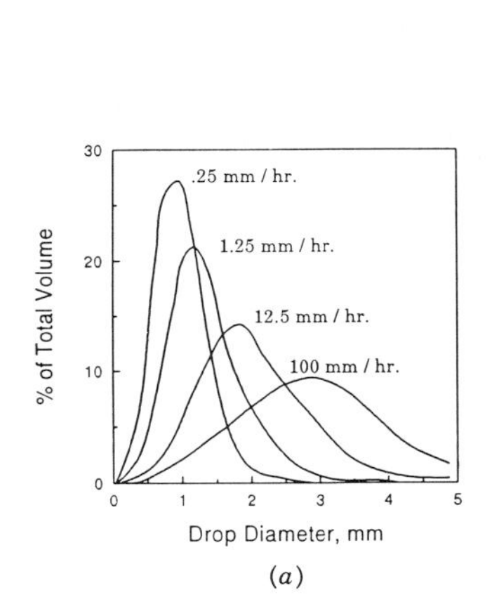

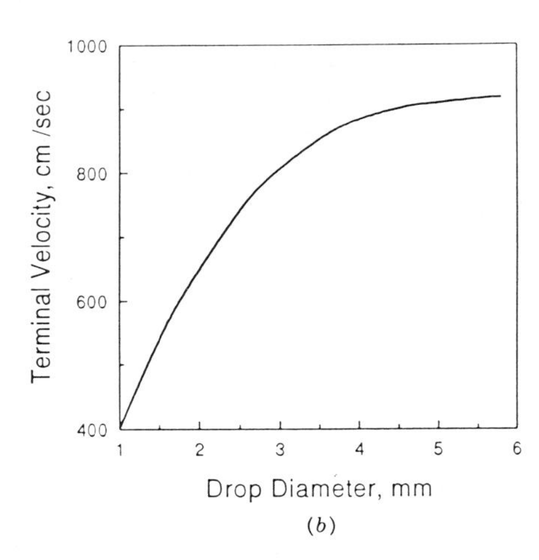

FIGURE 13.1. *Relationship between A. rainfall intensity and raindrop size, and B. raindrop terminal velocity and diameter.* A. From Laws & Parson, 1943; B. From Gunn and Kinzer, 1949. Republished by permission from (A) American Geophysical Union, (B) American Meteorological Society.

Slope Factors

Slope angle and length are important factors in determining the material eroded from an area (Fig. 13.2). Very high losses are possible from slopes above 12%. The above relationships raise questions about the stability of even moderately sloping landscapes with discontinuous vegetative cover. Areas with bunch-grass always have some bare soil, and debris dams and sediment sorting in micro-channels is common (Mitchell & Humphreys, 1987). Both the debris dams and micro-channels record run-off between the vegetated areas and the movement of some material. Instability is much easier to prove than stability.

Slope Shape Although we speak of hillslopes as one unit, they are complexes. Many shapes and an infinite number of relationships to other parts of the hillslope system can exist. There are few data on the effect of hillslope shape and position on erosion, although some existing models include these local relations (Moore et al., 1986).

Hillslopes which are cut into uniform materials have surface and subsurface water flow lines roughly parallel to surface contours (Selby, 1982, Fig. 5.16). Flow is diverging on nose slopes and converging on head and foot slopes. Run-off depth should be shallow on the convex nose slopes because run-off flow is diverging and the local drainage divide is nearby. The downslope contributing area increases much more rapidly in the concave head and foot slopes (O'Loughlin, 1986).

Surface erosion intensity depends upon the interactions among flow depth, raindrop size, surface slope and placement in relation to contributing areas. The most severely eroded areas are the nose slopes, steeper linear slopes, and some head or foot slope positions where flow concentrates (Fig. 13.3; Moore & Burch, 1986; Moore et al., 1986). Laboratory research shows that sediment movement is highest in the concave head slopes and lowest on the nose slopes (Schumm et al., 1987). Although sediment movement is highest on the head slopes, these positions also have thick mantles of hillslope sediment (Ruhe et

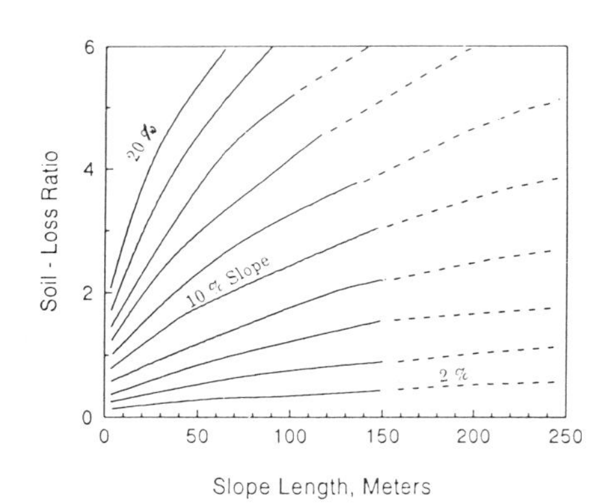

FIGURE 13.2. *Relationship between soil loss and slope steepness and length*. From Wischmeier and Smith, 1965.

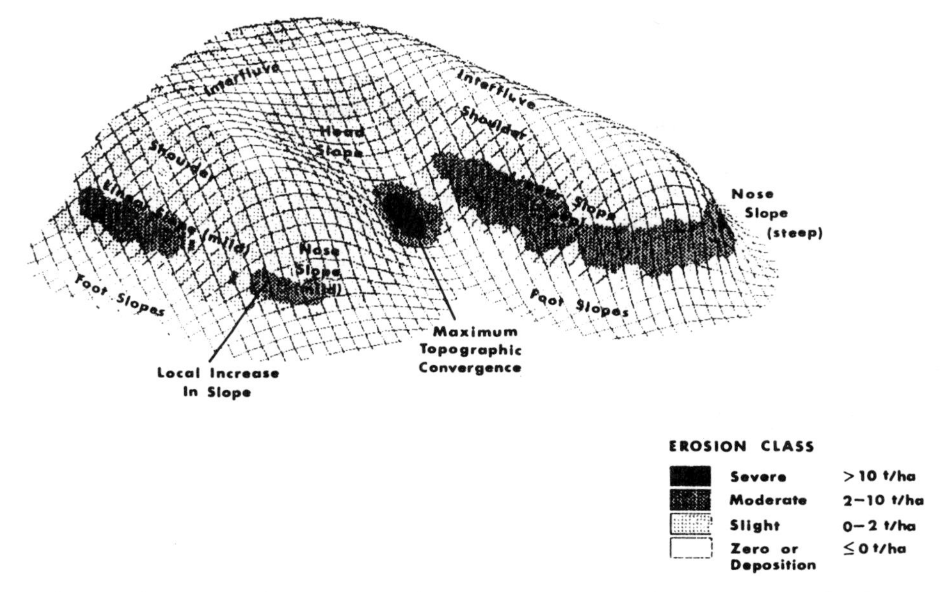

FIGURE 13.3. *Areas of maximum and minimum erosion in relation to topography*. From Moore et al. 1986, Fig. 1, p. 2374. Republished by permission from Soil Sci. Soc. Amer. and I.D. Moore.

al., 1967). Erosion removes the sediment mantle when a channel trenches headward into the head slope. Deposition reestablishes the sediment mantle when the adjacent slopes grade to the bottom of the new channel.

Visual evidence of erosion in cultivated fields usually occurs on the shoulder, linear and nose slopes. In eroded landscapes, these positions have more area of B horizon exposed than other parts of the landscape (Daniels et al., 1985). This observation supports conclusions of Moore and Burch (1986) who used stream power as a measure of erosion potential.

Interpretation of the visual evidence for shoulders and nose slopes should be very cautious. These positions are the driest areas of the landscape (diverging flow), and it is probable that they always produced less biomass and developed thin surface horizons. The data of Schumm et al. (1987) may have more validity in cultivated fields than parts of the Moore et al. (1986) model.

Computer Modeling of Slope Changes. Meyer and Kramer (1969) used the Universal Soil Loss Equation (USLE) to predict how slope shape changes when eroded. Their computer model used uniform two-dimensional, concave, convex and complex (upper convex, lower concave) slope shapes with mean gradients of 5 and 10 percent. All slopes had a 20-foot elevation difference between the top and the bottom. Coefficients allowed each erosion period a soil loss of about 40 tons per acre on a 5% slope 400 feet long. A flat area was beyond the toe of the slope.

The simulation data show that the concave slope had the least erosion, and the convex slope the largest amount (Fig. 13.4). Both the concave and complex slopes had more erosion at the shorter slope lengths than did the convex or uniform slopes. But the total volume removed was less. Also the concave and complex slopes had areas of deposition on the lower parts, the area of decreasing sediment load (Fig. 13.4). The 10% slope changed gradient twice as fast as the 5% slopes, but sediment loads were similar at equivalent downslope elevations. After 50 erosion periods, the concave and complex slopes retained their initial forms. The uniform and convex slope developed convex-concave forms after 50 erosion periods (Fig. 13.5).

Under the stable base-level of this simulation, deposition occurred on the concave part of the slope, and erosion was greatest at mid-slope. A diagram of landscape erosion (Fig. 13.3; Moore et al., 1986) locates the moderate and

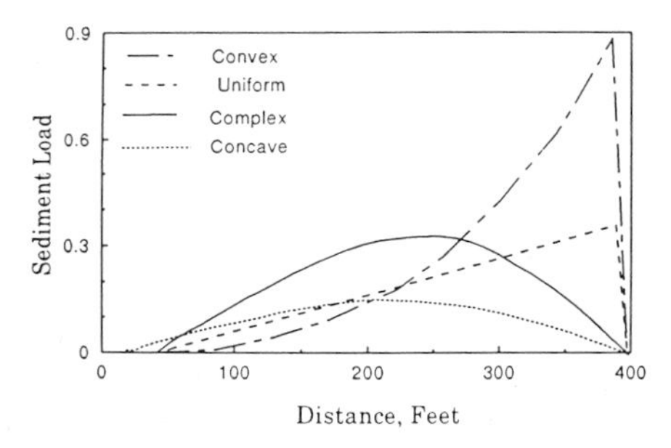

FIGURE 13.4. *Relationship between slope length, form and sediment load.* Redrawn from Meyer and Kramer, 1969, Fig. 2, p. 523. Republished by permission from American Society of Agricultural Engineers.

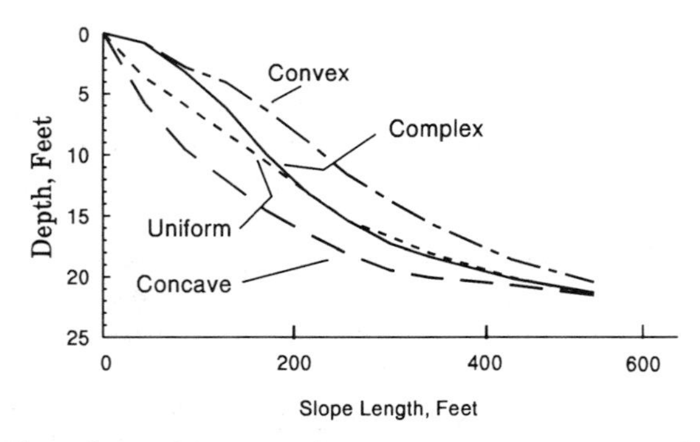

FIGURE 13.5. *Slope shape after 50 periods of erosion.* Redrawn from Meyer and Kramer, 1969, Fig 3, p. 523. Republished by permission from American Society of Agricultural Engineers.

severe erosion classes at mid-slope. But the zone of maximum erosion over a long period probably occurs above mid-slope if the valley walls recede as Ruhe's model suggests.

Predicting the zone of maximum slope erosion is difficult because few established criteria are available. Simulations or even actual measurements under field conditions usually do not consider the long-term effect of erosion and deposition under a rising base-level. A rising base-level forces the zone of maximum erosion to move upslope gradually. This means that the higher slopes undergo more total erosion than the lower slopes.

Problems in Quantifying Hillslope Evolution

The quantitative data developed help explain some processes operating on the landscape. We still do not have conclusive evidence of how a landscape evolves over time. The assumptions made limit the predictive model's scope. Many modelers have a narrow or incomplete appreciation of field conditions. Testing their models is difficult because landscapes require considerable time to develop. Most modelers lack field experience, so it is difficult for them to test their products, but most experienced field personnel are not good modelers.

Cotta's 1816 observation that "the forester who practices much writes but little, and he who writes much practices but little" (1902) applies to modern day earth scientists. We need teams of modelers and experienced field personnel, but questions will remain.

Similar landscapes can develop in several different ways depending upon the controlling variables. The environmental combinations that control a landscape are almost infinite. We will learn much from a continuing effort, although we may never reach our goal of quantifying how a landscape develops.

MASS WASTING

Mass wasting is the downslope movement of soil or rock material under the influence of gravity. Mass movement does not directly involve water, air or ice, although they frequently reduce the strength of slope materials. Table 13.2 is one of several classifications of mass wasting.

Types

The three major types of mass movement are heave, slide and flow. Rockfall may be a probable mass movement mechanism. Heave is change from heating and cooling, wetting and drying, freezing and thawing, and water pressure from periodic saturation. Flow velocity decreases with depth, but velocity remains constant with depth in slides (Fig. 13.6).

TABLE 13.2. Rate of Mass Movement Processes

Rate	Ice	Rock or Soil	Water
		Flow	
Imperceptible	Solifluction	Creep	Solifluction
Slow to rapid	Debris avalanche		Earth flow, Mud flow, Debris avalanche
		Slide	
Slow to rapid		Slump, Debris-slide, Debris-fall, Rockslide, Rock fall	

Modified from Sharpe, (1938), pp. 22, 50 and 65. Republished by permission from Columbia University Press.

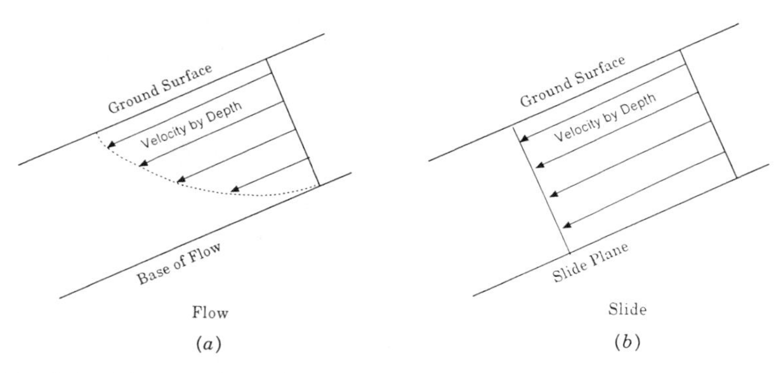

FIGURE 13.6. *Velocity with depth in flows and slides.* Redrawn from Small and Clark, 1982, Fig. 3.7, p. 35. Republished by permission from University of Cambridge Press.

Soil Creep Soil creep probably is a very understated process on moderate to strong slopes. For example, frost heave moves particles downslope by lifting them normal to the slope and then redepositing them during icemelt (Fig. 13.7). Frost heave is largely responsible for the surface "mulch" of soil material that obscures road cuts in late winter. Theoretically, creep is a heaving process followed by settling of the particles downslope (Fig. 13.7). Actual movement is erratic in time and space (Young, 1960). Young (1975) measured a linear downslope component of 0.25 mm/yr movement downslope and 0.31 mm/yr perpendicular to the ground. Rates are 0.1 to 15 mm/yr in vegetated soil and 0.5 m/yr in cold region talus.

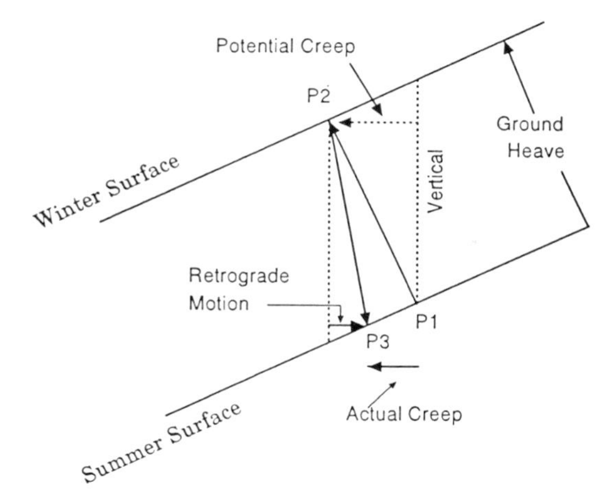

FIGURE 13.7. *Movement by frost creep.* Redrawn from Small and Clark, 1982, Fig. 3.3, p. 29. Republished by permission from University of Cambridge Press.

Terracettes Selby (1982) believes terracettes, small stair-step-like features on slopes, develop from soil creep. Each terracette, called "cat steps" and "cow paths" by some, has a nearly level area or tread and a sloping or nearly vertical riser. These features are common on steep slopes. Selby (1982) reported they cover 40% on some mudstone hillslopes. Terracettes are common in pastures on steep slopes, but they also form in forest. Some authors believe these features form by slumping, others (Brice, 1958) that they are erosional features in loessial landscapes.

Field evidence of surface creep is not always conclusive. It includes soil accumulation on the upslope side of trees and other obstructions, curved tree trunks, tilted walls, cross-slope cracks and rolls of turf (Selby, 1982). Others have used beds of saprolite bending downslope as evidence of creep (Graham, 1986). Although creep velocity is slow (Fig. 13.8), there is an overlap in flow rate between creep and solifluction.

Solifluction Solifluction is movement of saturated materials in periglacial areas. High moisture content and an impermeable layer of permafrost beneath the nearly fluid overlying material cause downslope movement. The movement contorts the sediments. The major differences between solifluction and creep are the special conditions required for solifluction and the slightly higher flow velocity.

Earth Flows and Mud Flows Slumps have curved failure planes and involve rotational movement of the soil mass. The area between the headwall and the backward rotating soil mass may have sag ponds or depressions. The failure has a stepped surface that merges downslope into a flow. The headwall may

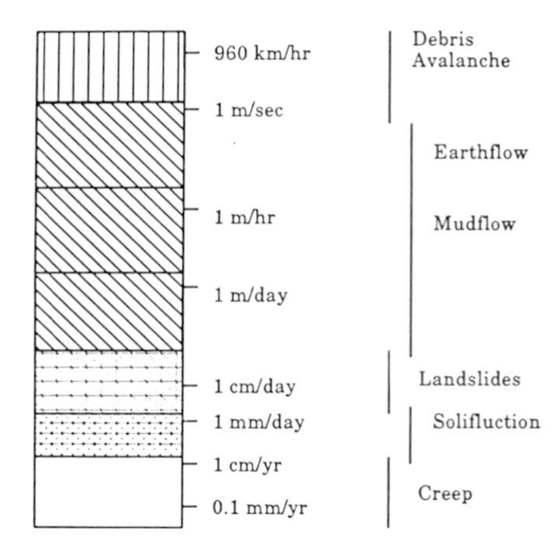

FIGURE 13.8. *Velocity of mass movement processes*. Redrawn from Small and Clark, 1982, Fig. 3.13, p. 42. Republished by permission from University of Cambridge Press.

continue to move upslope. Translational slides are common in soils. They may start as a coherent slide and change to debris slide as deformation and water content increase downslope. These slides may be thoroughly liquified flows at the base. Most translational slides occur during heavy rains. Figure 13.9 is a sketch of a landslide.

Debris flows are mixtures of granular solids with minor amounts of clay, entrained water and air. Debris flows can move on low slopes (Johnson & Rodine, 1984). This type of flow has the properties of debris avalanches, debris flows and mud flows defined by Varnes (1978). Debris slides, mud flows, rocky mud flows, mud slides, earth flows, mudspates, and lahars are materials deposited by similar processes (Johnson & Rodine, 1984). Debris flows have very coarse materials, but they flow as wet concrete and can fill houses without destroying the walls (Johnson & Rodine, 1984).

Debris flows are coarser and have higher velocity than mud flows. The snout is very coarse, with the following waves composed of finer debris. Figure 13.10 is an idealized diagram of a debris-flow arm showing the deposits formed by successive waves of debris. Where material is available, a debris flow can have very coarse bouldery levees (Fig. 13.11) with little or no finer material remaining in the channel.

Debris avalanches are very rapid events (Fig. 13.8) that can move large amounts of material for several kilometers (Small & Clark, 1982). Williams and Guy (1973) give excellent descriptions of avalanche deposits and processes. Avalanches are very destructive. Avalanches produced 3.2 to 4.6 million cubic feet of sediment per square mile in Nelson County, Virginia (Williams & Guy, 1973).

Several factors contribute to mass movement in soil materials (Sidel et al., 1985). Movement is common in steep topography with periodically saturated weak materials. Hurricanes and moderate intensity storms of long duration can move large amounts of material (Gryta & Bartholomew, 1989; Jacobson et al., 1989; Williams & Guy, 1973). Forest harvest and associated road construc-

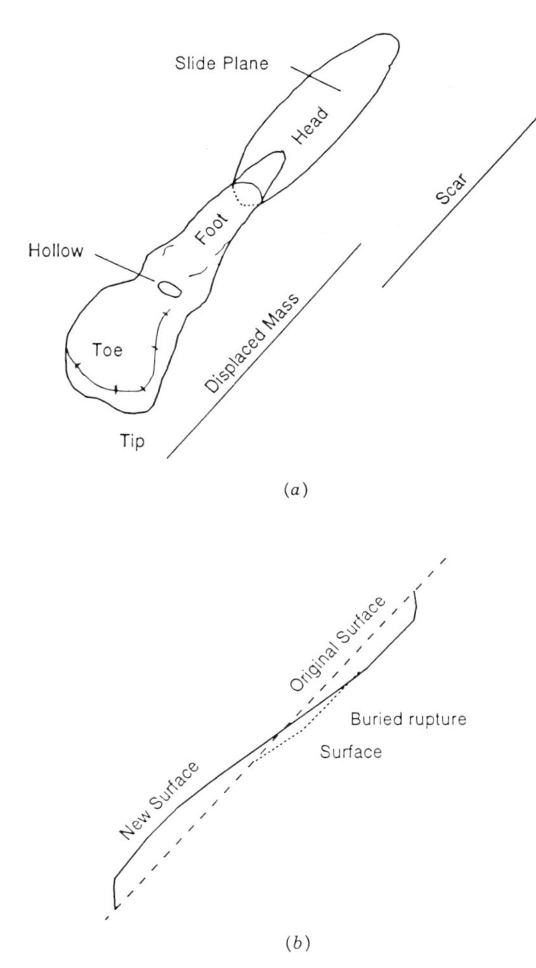

FIGURE 13.9. *Schematic of a landslide.* Redrawn after Crozier, 1973, Fig. 1, p. 81. Republished by permission from Gebruder Borntraeger.

tion increase the possibility of mass failure. The loss of root strength after forest harvest is a major factor in the increased failure of steep slopes (Sidel et al., 1985). Improper road placement and construction can undercut a slope. Mass movement frequently begins on the upslope wall of the roadbank. Areas with seismic activity are especially prone to the more dramatic forms of mass movement.

Summary

The obvious effects of landslides, mud flows and debris flows on steep slopes results in considerable study of these features. The effects of soil creep usually

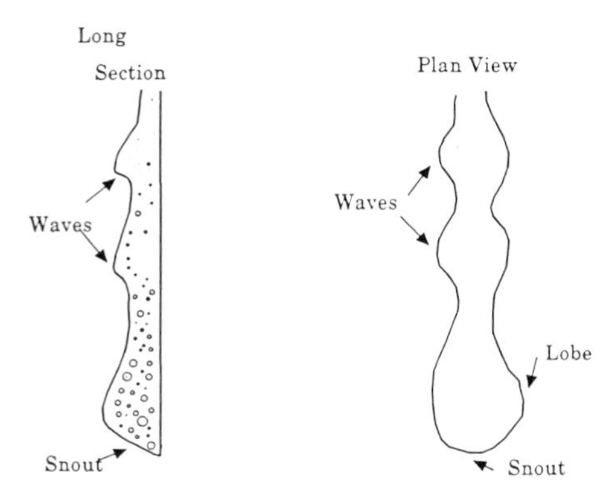

FIGURE 13.10. *Idealized schematic of a debris-flow arm* showing the deposits and general textural changes formed by successive waves of debris. Redrawn from Johnson and Rodine, 1984, Fig. 8.4, p. 263. Republished by permission from John Wiley & Sons.

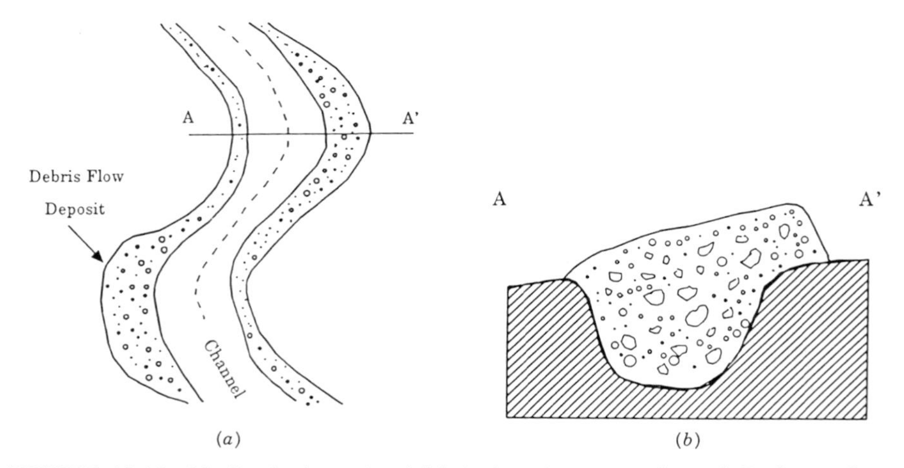

FIGURE 13.11. *Idealized schematic of debris deposits next to channel.* Redrawn from Johnson and Rodine, 1984, Fig. 8.22, p. 307. Republished by permission from John Wiley & Sons.

are not so obvious, and measurements, especially on gentle slopes, are difficult. Most of the reported studies are from slopes of 15 to 20 degrees. This makes for difficult extrapolation of results to agricultural soils on slopes of 2 to 12. Soil creep may be a major factor in shaping the landscape and moving soil materials downslope. Soil creep is possible with each wetting and drying cycle of swelling clays. Most of our attention is on the erosional aspects of landscape development and creep may play a major role. On most slopes, one must question whether the upper horizon of the soil formed in place or has moved downslope by creep and related processes.

Mud flows, slides and debris avalanches are common features in steeper areas in humid climates. The area affected by one storm can be large, and makes one question the Pleistocene age given to many of these deposits. For example, a two-day rainfall of moderate intensity triggered more than 1,000 slope movements in a 1,040 km^2 area of West Virginia and Virginia (Jacobson et al., 1989). Most of the movement was shallow slips and slip-flows in thin colluvium and residuum on shale slopes. Several large debris avalanches occurred in sandstone on slopes of 30 to 35 degrees. Intense rainfalls, such as the 1969 hurricane Camille, generate debris avalanches in the eastern United States (Gryta & Bartholomew, 1989).

Within Nelson County, Virginia, most of the debris avalanche activity from Camille was within the area of >40 cm rainfall (Gryta & Bartholomew, 1989). Avalanche density within this area was from 0.1 to 6 percent of the total area. There is no direct relation between landslide density and rainfall or rock-type. The high landslide density within areas of biotite-rich rocks most likely reflects topography. Rainfall, topography and lithology apparently control about 88% of avalanche chutes, and rock fabric controls about 10%.

Williams and Guy (1973) described the erosion and landscape changes produced by hurricane Camille in a small area of Virginia. They found the avalanche scar usually extends from the top of the slope to the region where the local hillslope gradient was steepest. The typical chute scoured was 61 to 244 m long, 8 to 23 m wide and 0.3 to 1 m deep. Slopes steeper than 27% are most susceptible to failure (Peatross, 1986). Many debris avalanches move down existing groves, minor channels or depressions on the hillslope. The north, northeast and east slopes have more debris avalanching than slopes with other directions. The deposits associated with debris avalanching are:

1. rare deposits near the base of hillslopes,
2. mountain channel deposits, sometimes debris piles behind log jams,
3. alluvial fans in intermontane valleys,
4. deltas where highway embankments temporarily blocked streams,
5. flood-plain sediments.

Williams and Guy (1973) found that nearly all material larger than fine sand remains within 5 to 10 miles of the source. They also reported that slightly more than half the erosion in the area was from enlargement of stream channels.

Materials deposited by various mass movement processes vary widely in lithology and particle size. Slumps and minor flows have little sorting. Sorting of debris avalanche deposits varies with transport distance and the amount of later reworking by water (Williams & Guy, 1973). We must use considerable caution when we interpret soil genesis in areas subject to mass movement. Minor slumps and flows that have minimal sorting may retain the B horizon color and development of "old" deposits and soils. Weathering indices may be of little value in establishing the approximate time of stabilization because these properties are from earlier periods of weathering. Table 13.3 gives the

TABLE 13.3. Features Indicating Active and Inactive Landslides

Active	Inactive
Scarps, terraces and crevices with sharp edges.	Scarps, terraces and crevices with rounded edges.
Crevices and depressions without secondary infilling.	Crevices and depressions infilled with secondary deposits.
Secondary mass movement on scarp faces.	No secondary mass movement on scarp faces.
Surface-of-rupture and marginal shear planes show fresh slickensides and striations.	Surface-of-rupture and marginal shear planes show old or no slickensides and striations.
Fresh fractured surfaces on blocks.	Weathering on fractured surfaces of blocks.
Drainage system disarranged; many ponds and undrained depressions.	Integrated drainage system.
Pressure ridges in contact with slide margin.	Marginal fissures and abandoned levees.
No soil development on exposed surface-of-rupture.	Soil development on exposed surface-of-rupture.
Fast-growing spp.	Slow-growing vegetation spp.
Different vegetation on and off slide.	No distinction between vegetation on and off slide.
Tilted trees with no new vertical growth.	Tilted trees with new vertical above inclined trunk.
No new supportive, secondary tissue on trunks.	New supportive, secondary tissue on trunks.

From Crozier, 1984, Table 4.10, p. 137. Republished by permission from John Wiley & Sons.

features that suggest active and inactive landslides. The properties given are useful field criteria.

The effect of a single storm such as Camille suggests that erosion modified most mountain slopes in the eastern United States since Pleistocene. Creep alone probably moved large volumes of material to the stream system during the last 10,000 years. Dating the deposits and applying this information to soil genesis studies is not easy.

ADDITIONAL READING

Brundsden, D. and D.B. Prior. (Eds.) (1984). *Slope Instability*. New York: Wiley.

Schultz, A.P. and C.S. Southworth. (Eds.) (1987). *Landslides of Eastern North America*. U.S. Geol. Survey Circular 1008.

Schuster, R.L. and R.J. Krisek. (Eds.) (1987). *Landslides, Analysis and Control*. Nat. Acad. Sci., Transportation Research Board Spec. Rpt. 176.

REFERENCES

Alvarado, A., C.W. Berish, and F. Peralta. (1981). *Soil Sci. Soc. Am. J.*, 45:790–794.

Black, T.A. and D.R. Montgomery. (1991). *Earth Surface Processes and Landforms*, 16:163–172.

Bonell, M., R.J. Coventry, and J.A. Holt. (1986). *Catena*, 13:11–28.

Brice, J.C. (1958). *U.S. Geol. Survey Bull.*, 25C.

Burras, C.L. and W.H. Scholtes. (1987). *Soil Sci. Soc. Amer. J.*, 51:1541–1547.

Ciolkosz, E.J., W.J. Waltman, T.W. Simpson, and R.R. Dobos. (1989). *Geomorphology*, 2:285–302.

Cleaves, E.T., A.E. Godfrey, and O.P. Bricker. (1974). *Geol. Soc. Am. Bull.*, 81:3015–3032.

Cotta, H. (1902). *Forest Quarterly*, 1:3–5.

Crozier, M.J. (1973). *Zeitschrift fur Geomorphologie*, 17:78–101.

Crozier, M.J. (1984). Field assessment of slope instability. In *Slope Instability* (pp. 103–142). Ed. by D. Brunsden and D.B. Prior. New York: Wiley.

Daniels, R.B., J.W. Gilliam, D.K. Cassel, and L.A. Nelson. (1985). *Soil Sci. Soc. Amer. J.*, 49:991–995.

Denny, C.S. and J.C. Goodlet. (1956). *Microrelief Resulting from Fallen Trees*. U.S. Geol. Survey Prof. Paper 288:59–66.

Denny, C.S. and W.H. Lyford. (1963). *Surficial Geology and Soils of the Elmira-Williamsport Region, New York and Pennsylvania*. U.S. Geol. Survey Prof. Paper 379.

Finlayson, B. and I. Statham. (1980). *Hillslope Analysis*. Boston: Butterworths.

Graham, R.C. (1986). *Geomorphology, Mineral Weathering and Pedogenesis in an Area of the Blue Ridge, North Carolina*. Unpublished doctoral dissertation. Raleigh: North Carolina State Univ.

Gryta, J.J. and M. J. Bartholomew. (1989). Factors influencing the distribution of debris avalanches associated with the 1969 Hurricane Camille in Nelson County, Virginia. In *Landslide Process of the Eastern United States and Puerto Rico* (pp. 15–28). Ed. by A.P. Schultz and R.W. Jibson. Geological Society of America Special Paper 236.

Gunn, R. and G.D. Kinzer. (1949). *J. of Meteorology*, 6:243–248.

Hole, F.D. (1982). Effects of animals on soils. *Geoderma*, 25:75–112.

Hugie, V.K. and H.B. Passey. (1963). *Soil Sci. Soc. Am. Proc.*, 27:78–82.

Jacobson, R.B., E.D. Cron, and J.P. McGeehin. (1989). Slope movements triggered by heavy rainfall, November 3-5, 1985, in Virginia and West Virginia, U.S.A. In *Landslide Processes of the Eastern United States and Puerto Rico* (pp. 1–13). Ed. by A.P. Schultz and R.W. Jobson. Geological Society of America Special Paper 236.

Johnson, A.M. and J.R. Rodine. (1984). Debris flow. In *Slope Instability* (pp. 257–361). Ed. by D. Brunsden and D.B. Prior. New York: Wiley.

Laws, J.O. and D.A. Parsons. (1943). *Amer. Geophys. Union Trans.*, 24:452–460.

Lyford, W.H. (1963). *Harv. For. Pap.*, 7:1–18.

Meyer, D.L. and L.A. Kramer. (1969). *Agri. Eng.*, 50:522–523.

Mitchell, B.P. and G.S. Humphreys. (1987). *Geoderma*, 39:331–357.

Moore, I.D. and G.J. Burch. (1986). *Soil Sci. Soc. Am. J.*, 50:1294–1298.

Moore, I.D., G.J. Burch, and E.M. O'Loughlin. (1986). *Soil Sci. Soc. Am. J.*, 50:1374–1375.

O'Loughlin, E.M. (1986). *Water Resources Res.*, 22:794–804.

Pavich, J.J. (1986). Processes and rates of saprolite production and erosion of a foliated granitic rock of the Virginia Piedmont. In *Rates of Chemical Weathering of Rocks and Minerals* (pp. 551–590). Ed. by S.M. Colman and D.P. Dethier. Orlando, FL: Academic Press.

Peatross, J.L. (1986). *A Morphometric Study of Slope Stability Controls in Central Virginia.* Unpublished master's thesis. Charlottesville: University of Virginia.

Ruhe, R.V., R.B. Daniels, and J.G. Cady. (1967). *Landscape Evolution and Soil Formation in Southwestern Iowa.* USDA Tech. Bull., 1349.

Schafer, J.P. and J.H. Hartshorn. (1965). The quaternary of New England. In *The Quaternary of the United States* (pp. 113–128). Ed. by H.E. Wright. Princeton: Princeton Univ. Press.

Schumm, S.A., M.P. Mosley, and W.E. Weaver. (1987). *Experimental Fluvial Geomorphology.* New York: Wiley.

Selby, J.J. (1982). *Hillslope Materials and Processes.* Oxford: Oxford Univ. Press.

Sharpe, C.F.S. (1938). *Landslides and Related Phenomena.* New Jersey: Pageant.

Sidel, R.C., A.J. Pearce, and C.L. O'Loughlin. (1985). *Hillslope Stability and Land Use.* Amer. Geophys. Union, Water Res. Mono. 11.

Small, R.J. and M.J. Clark. (1982). *Slopes and Weathering.* Cambridge: Cambridge Univ. Press.

Thorp, J. (1949). *Sci. Mon.*, 68:180–191.

Varnes, D.J. (1978). Slope movement types and processes. In *Landslides, Analysis and Control* (pp. 11–13). Ed. by R.L. Schuster and R.J. Krisek. Nat. Acad. Sci., Transportation Research Board Spec. Rpt. 176.

Walker, P.H. (1966). *Postglacial Environments in Relation to Landscape and Soils on the Cary Drift, Iowa.* Ag. and Home Ec. Exp. Sta., Iowa State Univ. Res. Bull. 549:838–875.

Williams, G.P. and H.P. Guy. (1973). *Erosional and Depositional Aspects of Hurricane Camille in Virginia, 1969.* U.S. Geol. Survey Prof. Paper 804.

Wischmeier, W.H. and D.D. Smith. (1965). *Agr. Handbook 282.* Washington, DC: USDA.

Young, A. (1960). *Nature*, 188:120–122.

Young, A. (1975). *Slopes.* N.Y.: Longman.

Zaslavsky, D. and G. Sinai. (1981). *J. Hydraulics Div. ASCE*, 107:17–35.

14 Time and Soil Formation

INTRODUCTION

Most soil scientists have a strong bias about time and its influence on soil formation. Many papers discuss the subject and usually conclude that old soils have "stronger development" than young soils. Many factors other than age can influence soil morphology. Ideas about the influence of time on soil formation developed from the morphologic expression of soil often are incorrect. Some soil scientists think the soil is the same age as the deposit in which it forms. Thus, they conclude that old soils occur on old deposits. It is true that deeply weathered soils commonly occur on old deposits, but so do soils that are only slightly weathered. Deeply weathered soils can occur on limited areas of young geomorphic surfaces. The soils are in truncated relict profiles or weathering zones and inherit part of their morphology and weathering from another period of soil formation. By considering the weathering zone as part of the original geomorphic surface, we lose information about history of the younger soil. We then ignore the later erosion and concentrate upon soil morphology as an age indicator of the soil and surface.

SOIL MORPHOLOGY AS A TIME INDICATOR

Geomorphologists commonly use soil properties to help date deposits (Birkeland, 1984; Birkeland et al., 1991), but this method has many weaknesses. For soil morphologic evidence to be accurate, the sites used must be on constructional surfaces with the same microenvironment. Any correlations made using soils should be very local. Dating sediments from soil properties is risky because we know more about stratigraphy and geomorphology than we do about soil development. The time required to develop a particular degree of soil horizon expression depends upon microenvironment and material more than upon weathering duration.

A major problem with using soil morphology as an indicator of pedogenic time is that process intensity and soil material composition control the development. As Schumm (1977) has so clearly stated,

> Time here refers to the passage of time . . . and in itself has no influence on the landscape; rather it records the accomplishments of the system. (p. 19)

Time in the above context is important only to help establish a starting and stopping point and to compute process rate. The process may have started with material exposure to soil formation, or it may have begun late in the soil history. Time is a passive factor. Time only allows the accumulation of products of non-instantaneous processes.

To illustrate the passive nature of time, what kind of soil developmental sequence would we expect to find in the following deposits? The weathering periods are 100, 1,000, 10,000 and 40,000 years. Assume a nearly level but well-drained surface under a continuous vegetative cover with no erosion. The climate is humid, no aerosols fall upon the site, nor are other materials added to the surface.

1. A medium quartz sand with <3% fines.
2. A fine sandy loam with abundant mafic minerals and calcic plagioclase.

Which line (Fig. 14.1) would most closely fit A and B horizon development when the above periods end? An infinite number of lines probably would fit, depending upon seasonal rainfall distribution and temperature, and total rainfall.

Time—Relative, Effective and Absolute

Relative Time Stratigraphers and geomorphologists use relative time or age when data are not available to place a sediment, geomorphic surface or other earth features into an age sequence. Placing a feature into a relative time sequence by using stratigraphic and geomorphic evidence is a scientifically sound and accepted practice. What can one use in the soil profile that is an accurate indicator of relative age? Probably very little if anything, because one cannot hold all things constant. We can make relative comparisons of mineralogy, B horizon clay content and degree of horizonation. But we have

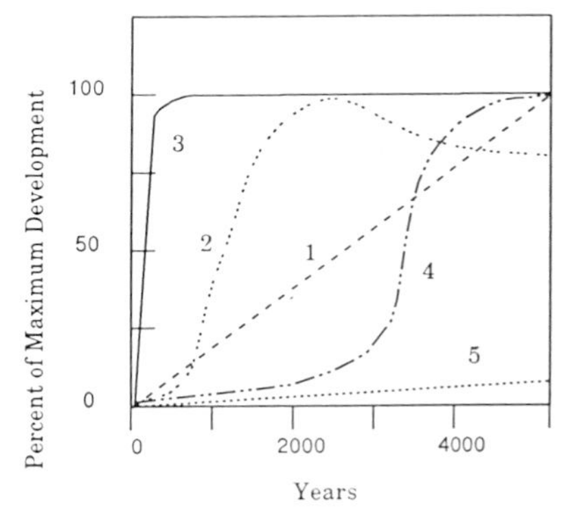

FIGURE 14.1. Possible shape of curves for A and B horizon development over time with quartz sand and mafic sandy loam.

no knowledge of the past weathering intensity, or whether the observed development is linear or curvilinear (Fig. 14.1).

Effective Time Smith (1942) used effective age to explain the progressive west to east changes of soil morphology in Illinois loess. His method of calculation is

- Loss of carbonate since close of loess deposition, 17.7 = 1 time unit.
- Loss of carbonate during loess deposition, 29.8 = 1.7 time units.

Using the above, Smith calculated the average age of the upper 30 inches of a 300 inch thick loess deposit as being equal to:

$$1.0 + 1/2 \text{ time required to deposit a tenth of the loess, or}$$

$$1.7/10 \times 1/2 = 0.1 \text{ or time} = 1.1.$$

$$\text{For a 30-inch loess deposit, } 1.0 + (1.7 \times 1/2) = 0.85 \text{ or } 1.85$$

By these calculations, the thinner loess weathered longer, or the effective time was 0.8 greater, than the thick loess. Thus, effective age explained the differences in soil morphology.

Considering the dating tools and ideas available when Smith completed his work, this is an interesting and well-documented approach. If present data are correct, Smith's approach placed too much emphasis upon the effect of time on soil development (Ruhe, 1969).

Modern evidence (Kleiss & Fehrenbacher, 1973) shows that the thick and thin loesses are the same age. All stable loess surfaces across the transect used by Smith are 12 to 14 thousand years old. There is no correlation between age and loess thickness. On the stable interfluves, the weathering time for soils is the same. Other factors must explain the observed differences in morphology.

Soil scientists commonly interpret soil age from soil morphological features. This approach ignores the parent material, depositional processes, relationships among geomorphic surfaces within landscapes, and changes in microenvironment within a geomorphic surface. Soils form from a variety of interrelated factors that may not be constant over geologic time. Interpreting the pathways and rates of soil genesis from morphological features alone is a risky venture.

Stone (1975) cautioned those who interpret soils for management practices when he stated that "the reciprocal influences of soil upon forest and forest upon soil are not easily disentangled." He cited Crowther's (1953) remarks that "statements . . . have come to be believed in for no better reason than that people have talked a great deal about them." These concerns apply to those who compare degrees of soil development by morphological features or taxonomic classification alone.

As an example, we cite the investigation of a developmental sequence of loess soils on a nearly level upland flat in Illinois (Fig. 14.2). This study illustrates the problems of using soil morphology and chemical properties as a measure of age or degree of development. Smeck and Runge (1971) studied phosphorus availability and distribution in an area 184 by 306 m. Their study involved three soils: a Mollic Albaqualf, and two "less well-developed" soils, an Aquic Argiudoll and a Typic Haplaquoll. Relief between the end members was <0.6 m. Total and available phosphorus (Fig. 14.3) and

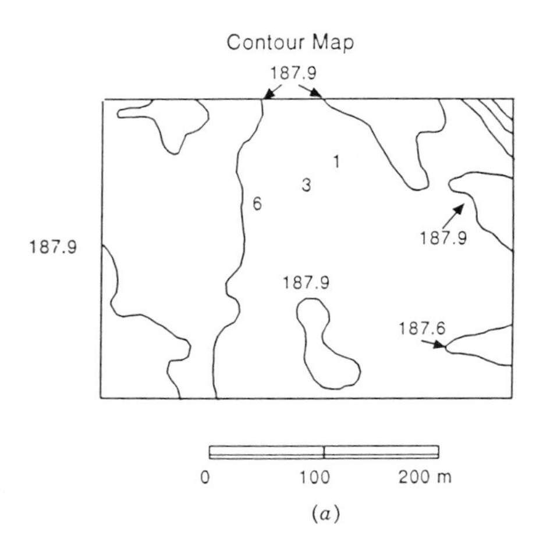

(*a*)

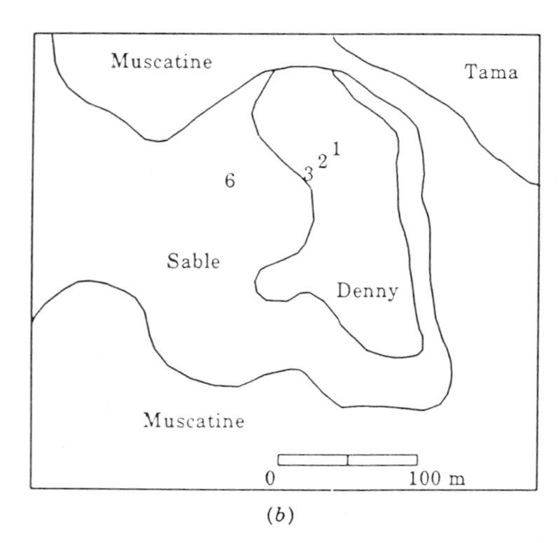

(*b*)

FIGURE 14.2. *Areal distribution of map units.* A. Contour map of study area. B. Soil map of study area. Redrawn from Smeck and Runge, 1971, Fig. 1, p. 953. Republished by permission from N.E. Smeck.

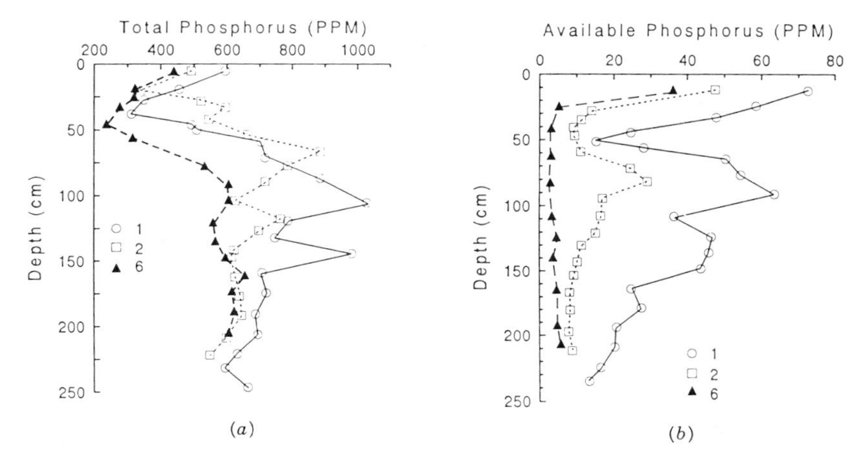

FIGURE 14.3. *Total (A) and available (B) phosphorus from a loess-derived developmental sequence of soils in Illinois.* Redrawn from Smeck and Runge, 1971, A. Fig. 3, p. 956, B. Fig. 5, p. 957. Republished by permission from N.E. Smeck.

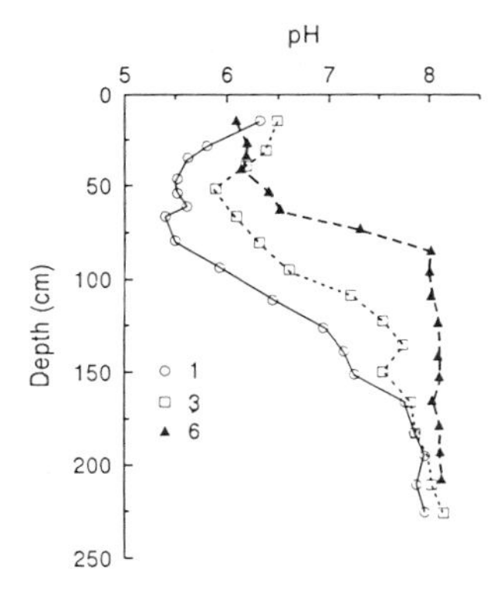

FIGURE 14.4. *pH changes across a developmental sequence.* Redrawn from Smeck and Runge, 1971, Fig. 4, p. 956. Republished by permission from N.E. Smeck.

pH (Fig. 14.4) show progressive changes from the Mollic Albaqualf to the adjacent Haplaquoll.

Data from Kleiss (1973) show a weathering interval of about 12,000 years. Effective age does not explain the differences shown. Degree of "soil development" is also suspect because it is difficult to quantify. If we say that some soils have stronger development than others, our basic assumptions are that the soils are from the same material and had the same environment. We also assume that weathering intensity and direction of soil formation are the same. The above assumptions probably are wrong in three of the four suppositions when applied to the Smeck and Runge (1971) data. The materials, ages of landscape and weathering duration are the same. The soil environment (hydrology), intensity and direction of soil formation are different, depending upon the site location within the hydrologic gradient.

We agree with Geiss et al. (1970) that soil development in these materials and climate is a result of the soil-water relationships occurring in the toposequence. Assignment of "degrees of development" to soils in the catena introduces a bias of perspective. All soil development is as strong as it can be under the imposed material, environmental, and topographic conditions. Interpreting soil development requires a great deal more information than can be gained from soil morphology and chemistry alone.

The paper by Knuteson et al. (1989) is an example of why we cannot use soil morphology to determine soil age. The study area was a recharge wetland in a landscape less than 9,000 years old. Soils with Bk and Bt horizons occur next to each other in the field (Fig. 14.5). The work clearly shows the effect of hydrology on the kind of soil developed. Only by studying the entire system—not just holes in the ground—can we partially understand the processes responsible for the soils within a landscape.

Soil Age

The age of a soil should refer to the time in years since exposure to subaerial weathering and soil formation. Use a radiometric dating method to determine soil age, if possible. If radiometric dates are not available, use other earth science criteria, not soil morphology. Often we cannot provide an absolute age for a soil or a process although we have accurate data on the duration of soil formation.

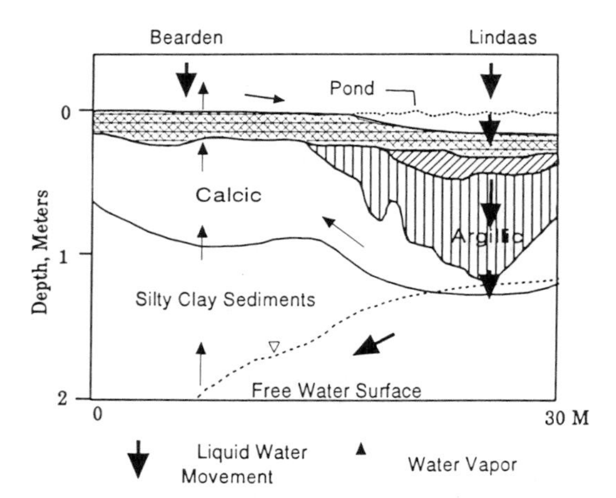

FIGURE 14.5. *Relationship of soils with calic and argillic horizons in and near a recharge wetland.* From Knuteson et al., (1989), Fig. 2, p. 496. Republished by permission from Soil Science Society America and J.A. Knuteson.

If we use the term *soil development*, it should refer *only* to horizon differentiation and not to an indication of age. Even more important than soil age is trying to understand when certain processes were active and what drove them. For example, does clay illuviation start immediately, or did it become active only in the later stages of soil development? When did landscape dissection and deep drainage make it possible for processes such as leaching and intense weathering to begin? Answers to the question of when a process started and stopped are as important as determining when a surface stabilized and soil formation began. And these questions are even more difficult to answer.

Even when we have accurate time measurements, we must use caution in their use and interpretation. Kleiss and Fehrenbacher (1973) date the base of the Wisconsinian loess at 25,000 years and the top at 12,000 years. This gives a uniform sedimentation rate of 0.34 cm/year of a loess section 4.42 m thick. They divide the loess into four zones and give each zone the following dates:

Base of zone I = 25,000 years
Base of zone II = 21,000 years
Base of zone III = 16,500 years
Base of zone IV = 15,000 and the top as 12,000.

Deposition rates at each site are as follows:

I & II	1.9 cm/yr
III	8.1 cm/yr
IV	5.4 cm/yr

Most sedimentation and erosional processes are not linear over time (Schumm, 1977). They do not form smooth curves unless they involve large periods, and then much of the needed data are missing. The sedimentation and erosional processes on a landscape are episodic; in other words, they go steady by jerks.

We believe the interest of soil scientists in time should be in determining when a soil process started or stopped. The total interval involved in profile development tells us little about process rate. We need to know the history of the study area in enough detail to develop chronograms such as those by Vreeken (1984). It is more important in genesis studies to know what started a process, or when the system reached a threshold (Schumm 1977). We need answers to the following questions:

- When and where did the process start?
- What was the rate and variability over time?
- What factor or factors control the process?

Obtaining the information for a chronogram is difficult in most areas, but it is a necessary first step.

REFERENCES

Birkeland, P.W. (1984). *Geoderma*, 34:115–134.

Birkeland, P.W., M.E. Berry, and D.K. Swanson. (1991). *Geology*, 19:281–283.

Crowther, E.M. (1953). *J. Soil Sci.*, 40:107–122.

Geiss, J.W., W.R. Boggess, and J.D. Alexander. (1970). *Soil Sci. Soc. Am. Proc.*, 34:105–111.

Kleiss, H.J. (1973). *Soil Sci.*, 115:194–198.

Kleiss, H.J. and J.B. Fehrenbacher. (1973). *Soil Sci. Soc. Am. Proc.*, 37:291–295.

Knuteson, J.A., J.L. Richardson, D.D. Patterson, and L. Prunty. (1989). *Soil Sci. Soc. Amer. J.*, 53:495–499.

Ruhe, R.V. (1969). *Quaternary Landscapes in Iowa*. Ames: Iowa State Univ. Press.

Schumm, S.A. (1977). *The Fluvial System*. New York: Wiley.

Smeck, N.E. and E.C.A. Runge. (1971). *Soil Sci. Soc. Amer. Proc.*, 35:952–959.

Smith, G.D. (1942). *Ill. Agri. Exp. Sta. Bull.*, 490:139–184.

Stone, E.L. (1975). *Philos. Trans. R. Cos. London*, Ser. B271:149–162.

Vreeken, W.J. (1984). *Geoderma*, 34:149–164.

15 Hydrology

INTRODUCTION

Many factors affect soil development, but none is more important than the abundance, flux, flow pathways, and seasonal distribution of water. Pedologists think of soils as reversed chromatographic columns with input at the top and outputs at the base. The input is water and the output is water plus dissolved and suspended constituents. Water in the soil column dissolves and transports materials. If concentrations are favorable, materials precipitate. Water not only adds and removes material from the soil, it rearranges the chemical and physical composition. Water can move in any direction depending on local conditions, and our first idea of downward flux may be incorrect.

Soil scientists, especially field personnel, know that water moves vertically and laterally in sloping landscapes. They also realize that soil structure and porosity greatly influence the rate and pathways of soil-water movement (Hursh & Fletcher, 1942). Scientists find it difficult to predict the amount and the rate of change of soil water.

The study of how and where soil water moves in the field is labor and time intensive. These studies include nearly all aspects of the hydrologic cycle that are difficult to measure. Understanding and predicting pathways and volumes of soil-water movement remain the primary challenges to wise use of soil and water resources.

The Dalrymple et al. (1968) land surface model (Fig. 10.4) shows vertical water movement through soil on interfluves. Water movement on slopes is both vertical and horizontal. Soil horizons change vertically, but water flow can be up, down, or lateral. Slope and restrictive horizons or beds control water flow direction. Climate controls water flow direction, amount, and rates (Gaskin et al., 1989; Timpson et al., 1986).

We must look at the effects of stratigraphy and geomorphology on site hydrology to understand soils. Abundant documentation supports this idea in the Prairie Pot Hole Region (Arndt & Richardson, 1988; Knuteson et al., 1989; Lissey, 1971; Miller et al., 1985). A general understanding of how water moves on and under the landscape is necessary before we present the complex relationships of how water movement influences soil development.

BASIC GROUNDWATER HYDROLOGY

Figure 15.1 is a schematic of the subdivisions of underground water made by hydrogeologists. The capillary fringe separates the unsaturated zone from the underlying saturated zone. The groundwater surface (water table) is the level that water stands in a bore hole. The saturated zone below the water table extends to depths reached by interconnected openings.

Figure 15.2 shows a groundwater system with unconfined and confined aquifers. A confining bed, or one that restricts water movement to and from the aquifer, overlies a confined aquifer (Heath, 1987).

Figure 15.3 shows some data needed to calculate groundwater flow direction in unconfined aquifers. Water movement in unconfined aquifers is from areas of high total head toward areas of low total head.

In Figure 15.3, water flows from well 1 toward well 2. The flow direction is in response to the difference in total head between the wells. The water table also slopes from well 1 to well 2 and shows the general direction of water movement. Determining actual direction of movement requires at least three wells open at the same level in an aquifer (Heath, 1987, p. 11).

The rate of groundwater flow in a hydraulically uniform aquifer depends upon the hydraulic gradient and the hydraulic conductivity of the aquifer. Hydrogeologists use Darcy's law to compute the quantity of flow:

$$Q = Ka\,(dh) - (dl) \qquad \textbf{(1)}$$

The quantity of water per unit time is Q; K is the hydraulic conductivity; A is the cross-sectional area at right angles to the flow direction; and dh/dl is the hydraulic gradient.

Hydraulic conductivity depends upon the size and continuity of the aquifer's pores and fractures and the hydraulic characteristics of the fluid

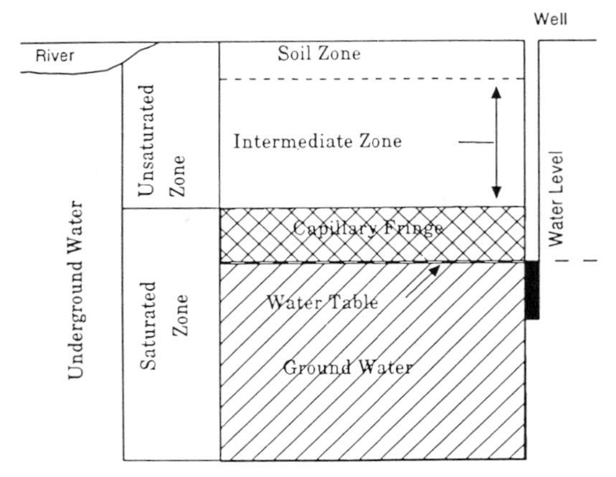

FIGURE 15.1. *Classification of underground water.* From Heath, 1987, p. 4.

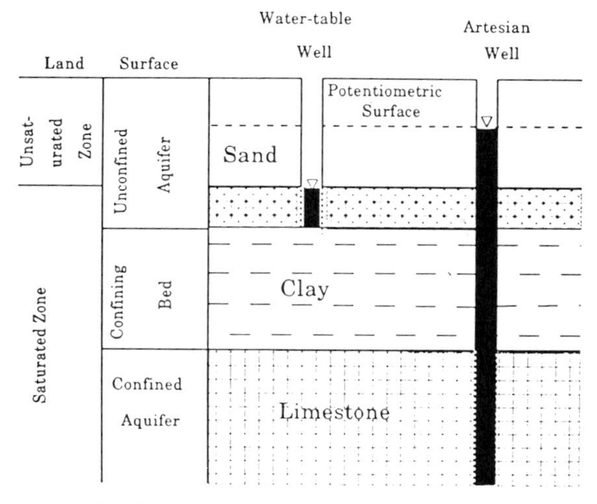

FIGURE 15.2. *Confined and unconfined aquifers*. From Heath, 1987, p. 6.

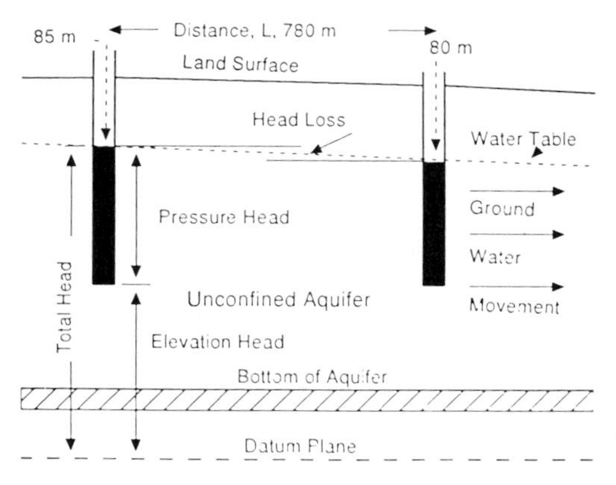

FIGURE 15.3. *A schematic illustrating total and pressure head between two wells in an unconfined aquifer*. From Heath, 1987, p. 10.

(Heath, 1987). Hydraulic conductivity units are velocity, or distance/time (equation 2).

$$K = \frac{Qdl}{Adh} = (m^3\, d^{-1})\,(m) - (m^2)\,(m) = \frac{m}{d} \tag{2}$$

One can compare the hydraulic conductivity of different materials and rocks by expressing hydraulic conductivity in terms of a unit gradient (Fig. 15.4). The ranges shown are extreme. Unfractured rock requires 1 million years to trans-

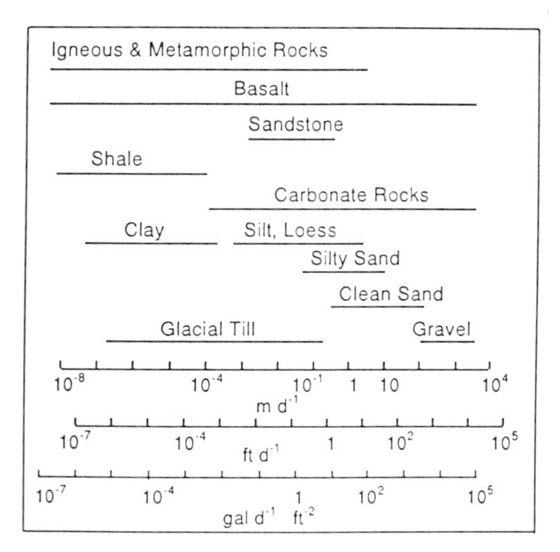

FIGURE 15.4. *Range in hydraulic conductivity of several kinds of rocks and sediments.* From Heath, 1987, p. 13.

mit one gallon of water through one square foot. Gravel can transmit 100,000 gallons of water per day through one square foot.

Figure 15.5 illustrates the time required for water reaching the water table in the recharge area to move through the groundwater system to the outlet. Water infiltrating at the watershed divide may take several centuries to move through the system. Sites close to outlets may require only days. Heath (1987) states that groundwater systems are more effective reservoirs than conduits. The residence time of water and properties of the aquifer are important in determining the dissolved load of groundwater in discharge areas. This dissolved load is an important consideration for soils in areas with salty aquifers.

Groundwater in an unconfined aquifer often is a subdued replica of the surface topography (Fig. 15.6). As a first approximation, in humid regions one can predict the depth to the water table and the flow direction from topography (Heath, 1987). This generalization is not always true. The complicated flow directions of the subhumid Prairie Pot Hole Region of North Dakota and Manitoba do not always follow the local topography. In this area, depressions may function as sites of recharge, flowthrough or discharge (Arndt & Richardson, 1988; Knuteson et al., 1989; Lissey, 1971). Mills and Zwarich (1986) give similar data for semiarid areas.

Flow Nets

Prediction of groundwater flow direction is possible from surface topography or water-table surface data. Only properly constructed flownets are proof of flow direction. Flownets have equipotential lines and flow lines. Equipotential lines connect points of equal head above a datum and represent the water-table surface (Heath, 1987). Flow lines are the idealized flow path of water as it moves through the aquifer. Water crosses equipotential lines at right angles.

Figures 15.7A and B show the equipotential and flow lines for a dissected landscape with a gaining stream. Figures 15.7C and D show the same lines for a losing stream. Figures 15.7A and C represent unconfined aquifers in humid and arid areas.

Groundwater recharge can move through a local, interrupted regional or regional flow system (Figs. 15.8A & B). Figure 15.8A is a schematic of a local flow system in landscapes with depressions at different elevations. Site A is at a higher elevation than site B, and both fill with water from spring snowmelt. The elevation changes between A and B drive the local flow pattern from A to B. Site A is a recharge depression (note the groundwater mound), and site B is a discharge unit.

Only part of the water in A flows to site B, some also moves to the regional and intermediate flow systems (Fig. 15.8B). Interrupted intermediate flow occurs where the topographic highs interrupt the decrease in elevation to the regional outlet. These interrupted and local flow systems may flow toward or away from the regional outlet. In the idealized system, the aquifer has uniform conductivity and infinite extent. Gradient alone controls the flow paths. Few natural systems contain conditions approximating the theoretical ideal.

The hydrologic characteristics and continuity of the underlying geologic material modify groundwater flow patterns. Figure 15.9 shows the equipoten-

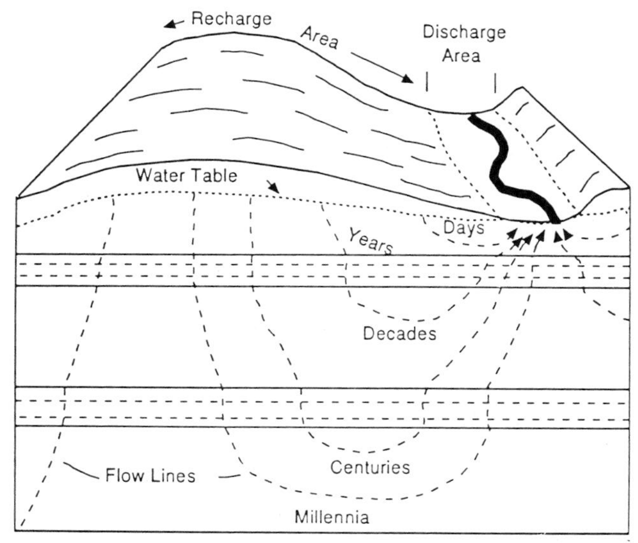

FIGURE 15.5. *Schematic of flow path and time required for water to move through a groundwater system.* From Heath, 1987, p. 14.

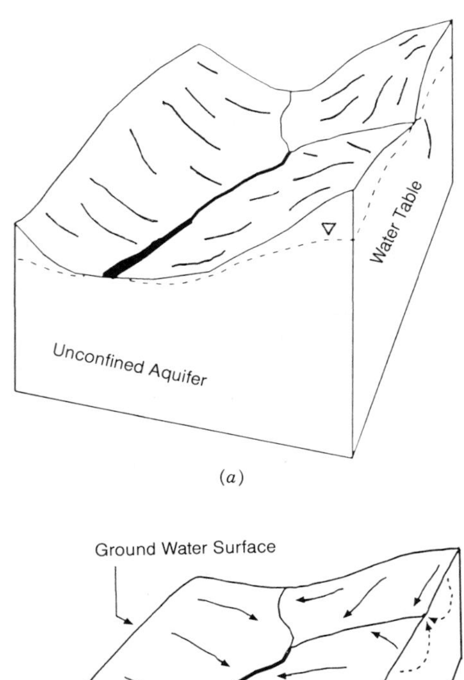

FIGURE 15.6. *Relationship of groundwater to topography*. From Heath, 1987, p. 20.

tial lines in an area with depressions and different underlying strata. Flow direction in Figure 15.9A is from the left-hand depression toward the one on the right. The highest depression is a recharge area, and the lowest (on the right) is a discharge area. The intermediate ponds with vertical equipotential lines are flowthrough ponds.

The shapes of the equipotential lines change when they cross layers of different hydraulic conductivity (Figs. 15.9B and C). This changes the volume of water moving to the discharge area. A large portion of water from the highest pond enters the sand strata and moves horizontally to some point beyond the diagram (Fig. 15.9B).

A larger portion of the water from the highest pond flows toward the discharge pond (right side of Fig. 15.9C) if a low permeability bed (confining bed)

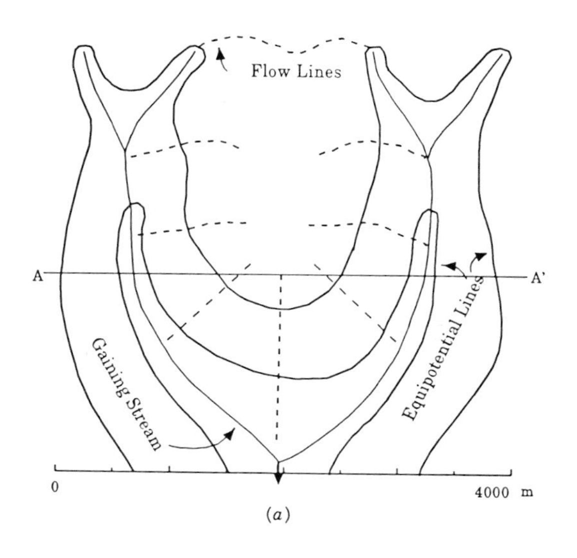

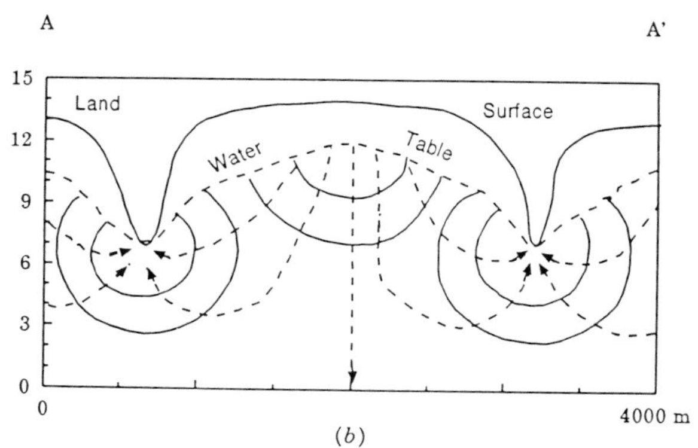

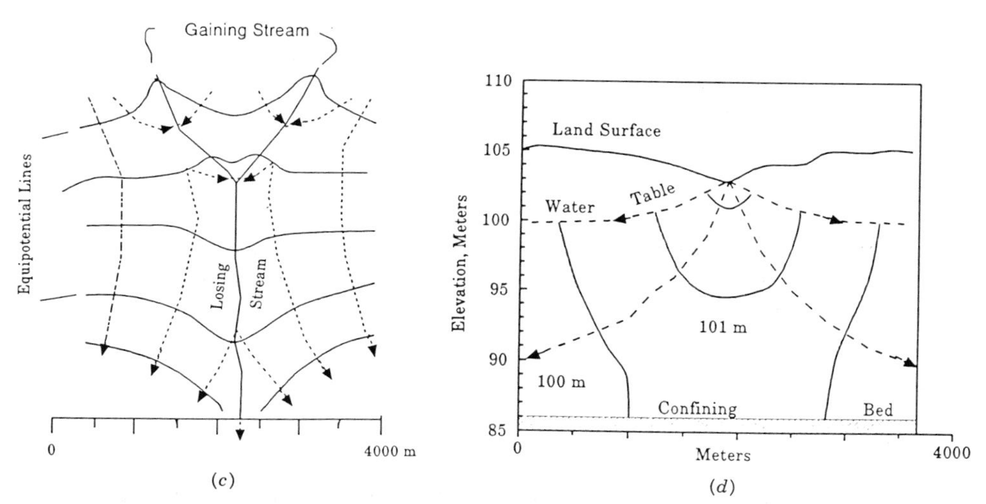

FIGURE 15.7. *Idealized equipotential and flow lines for a gaining and losing stream*. A. and B. Gaining stream. C. and D. Losing stream. From Heath, 1987, pp. 22 and 23.

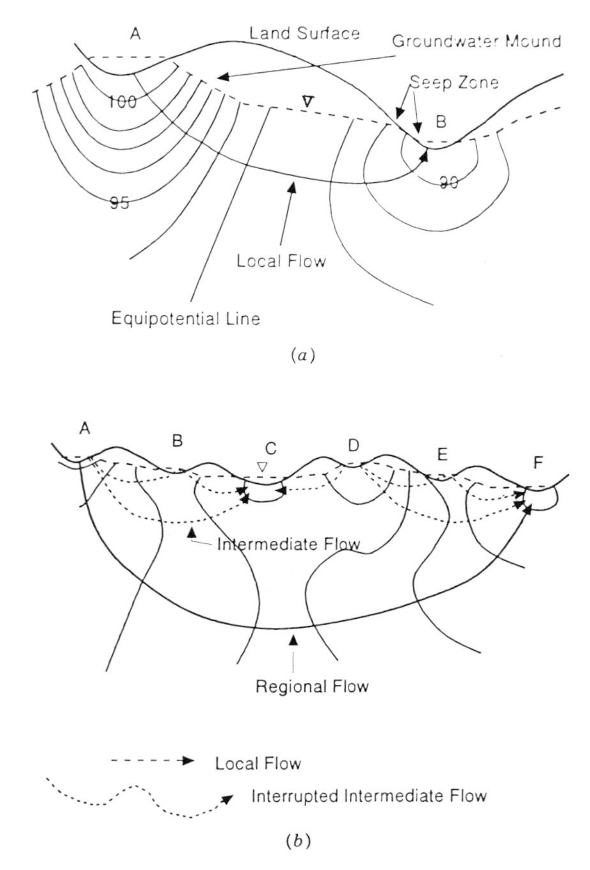

FIGURE 15.8. *Schematic of local, interrupted regional and regional groundwater flow systems.* A. Local flow. B. Regional and local flow. From Lissey, 1971, Fig. 3b, p. 336 & Fig. 6, p. 339. (A: Development of local flow system) (B: Development of intermediate flow system). Republished by permission from Geological Association of Canada.

occurs at depth. Permeability differences at shallow depths cause different temporal changes in the shallow flow system, whereas the deeper flow system may remain mostly unchanged (Lissey, 1971).

Complications

Idealized diagrams are useful in establishing general principles, but they may not accurately represent many field situations. The Prairie Pot Hole topography in North Dakota illustrates the difficulty in predicting subsurface water flow from surface form.

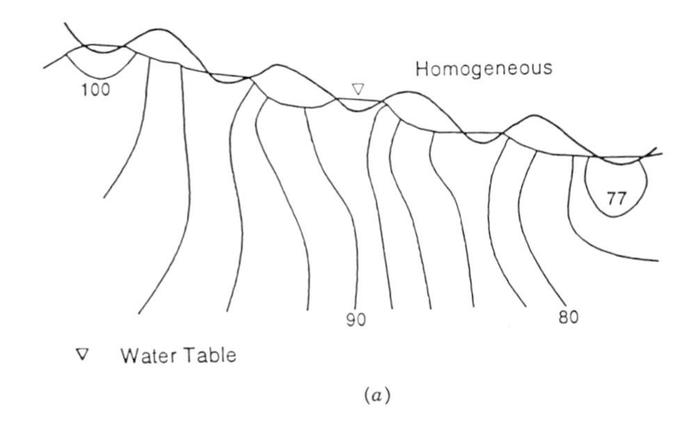

(*a*)

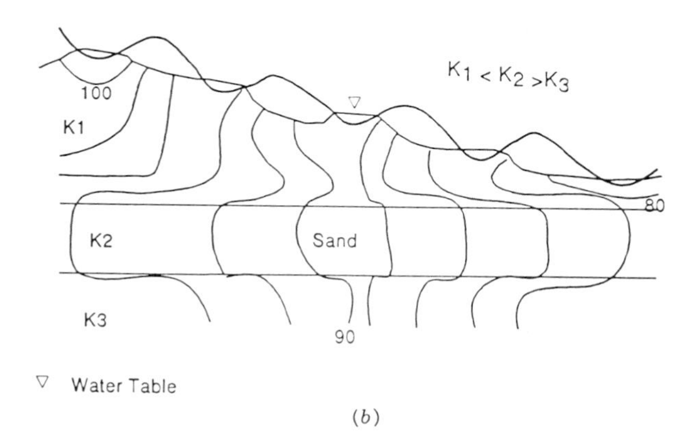

(*b*)

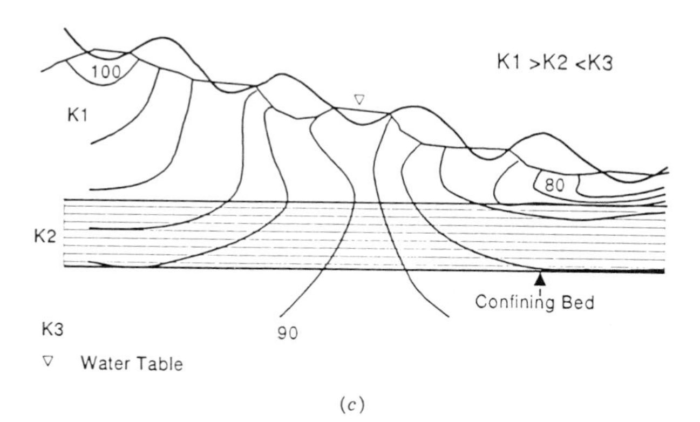

(*c*)

FIGURE 15.9. *Effect of permeability changes on flow patterns*. A. Homogeneous aquifer. B. Stratified with a deeper sand lens. C. Stratified with a deeper low-permeability bed. From Lissey, 1971, Fig. 7, p. 340. (Altered flow patterns due to permeability variations at depth). Republished by permission from Geological Association of Canada.

Figure 15.10 is a computer-generated generalized flownet of a local groundwater system in fractured till that lacks integrated drainage (Richardson et al., 1990). The condition shown occurs shortly after snowmelt when the depressions hold shallow ponded water. The equipotential lines beneath the recharge pond show that flow is downward and the depression is a recharge area. Note the water-table mound beneath the surface. Water infiltrating from the pond moves left, right or straight down.

Equipotential lines beneath the flowthrough pond are vertical, and groundwater flow is horizontal. Surface water enters one side and reenters the subsurface flow system on the other side. The flowthrough pond B has little or no vertical recharge to the groundwater system.

The curved equipotential lines beneath the discharge pond have the lowest value in the diagram. Flow in this area is upward into the depression, and groundwater discharges to the surface. The discharge pond should have surface water for longer periods than either the recharge or flowthrough pond.

Each surface pond or depression within Figure 15.10 has a different flow pattern. Position on the local landscape and the underlying materials combine to control the groundwater regime at each pond. Surface form and position help one predict flow patterns, but one should use these features only for the first estimate. Keller et al. (1988) and Richardson et al. (1990) believe conditions controlling recharge are complex, and the hydrologic significance of wetlands varies greatly. The permeability of the underlying materials produces much of the wetland variability.

Figure 15.11 gives the detail for a recharge, flowthrough and discharge pond on a local landscape. The closely spaced and decreasing values of the equipotential lines beneath the recharge pond (22) shows that water moves downward. Flow from pond 22 is downward and to the left, with some discharge on the right-side of pond 18. The shapes of the lines beneath pond 18 show discharge on the right-side and recharge on the left. Flow is upward into discharge pond 20, which receives water from ponds 22 and 18.

Figure 15.12 is a flownet from the North Carolina Middle Coastal Plain (Daniels et al., 1971). Sediments under the nearly level surface are sandy loam

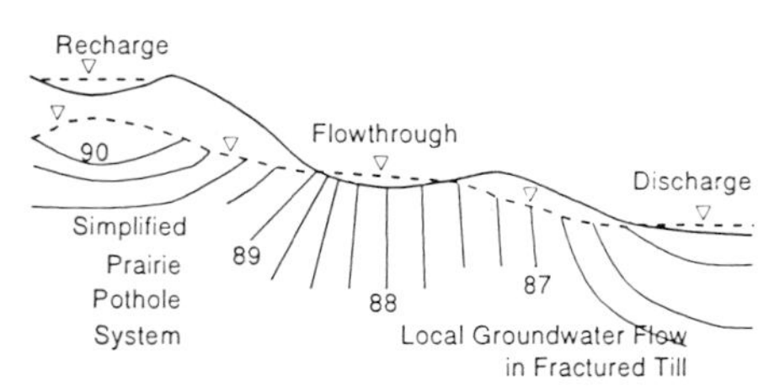

FIGURE 15.10. *Generalized flownet of a local groundwater system in fractured till that lacks integrated drainage.* From Richardson et al. (unpublished Ms, 1990), Fig. 1, p. 18. Republished by permission from J.L. Richardson.

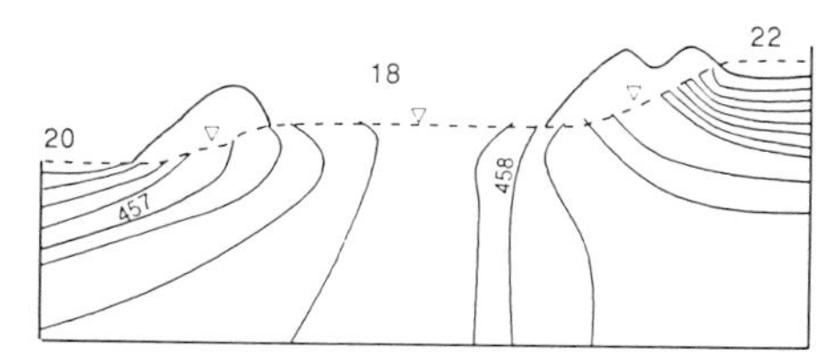

FIGURE 15.11. *Detail of expotential lines of a flownet* for a recharge pond (22), a flow-through pond (18) and a discharge pond (20). From Richardson et al. (unpublished Ms, 1990), Fig. 3a, p. 18. Republished by permission from J.L. Richardson.

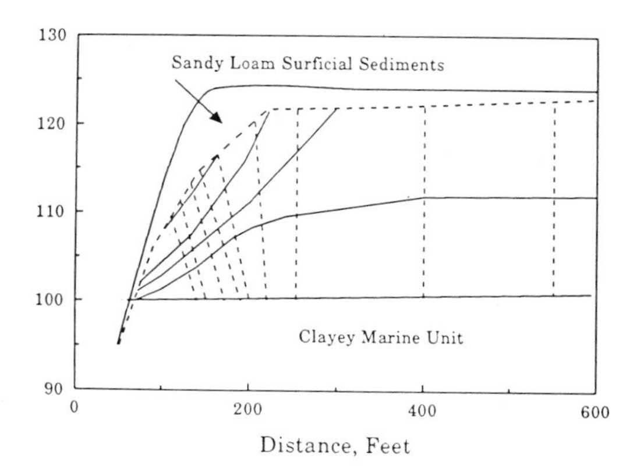

FIGURE 15.12. *Flownet of a Coastal Plain flat near an escarpment.* Original data from Daniels et al. (1971), Drawing from Richardson, et al. (unpublished Ms, 1990), Fig. 6, p. 18. Republished by permission from J.L. Richardson.

or coarser. The unconfined surface aquifer is about 12 m thick and overlies a thick clayey confining bed. Equipotential lines are vertical from the divide center to near the valley slope, proof that most of the flow is horizontal. Near the edge of the level surface, the curved equipotential lines show that water movement is downward toward the outlet above the confining bed.

NEAR-SURFACE HYDROLOGY AND SOILS

Flownet analysis is of little use to soil scientists unless relationships exist among soil morphology, soil chemistry and groundwater movement. How water moves in the subsurface has a direct effect upon soil morphology and chemical composition. The North Dakota Prairie Pot Hole Region clearly illustrates how site hydrology influences soil properties. These examples apply to almost any area that has at least short periods of excess water above evapotranspiration. The following paragraphs give some details of how site hydrology modifies soil morphology and chemical properties.

Figure 15.10 illustrates how water moves from a recharge, to a flowthrough and then to a discharge pond. Water entering the recharge pond is fresh, but dissolves some salt as it moves through the fractured till to the flowthrough pond. The flowthrough wetland, 18 in Figure 15.11, receives brackish water. From the flowthrough pond to discharge wetland 20, groundwater dissolves more salt and becomes saline (Richardson et al., 1990).

Water moves from the surface of recharge ponds (Fig. 15.10) downward to the water-table surface. Soils in recharge ponds have argillic horizons. Clay translocation probably occurs during the initial downward flux of water during the wet-up period (Thorpe et al., 1957). These recharge ponds usually have water only during the spring run-off periods, so clay movement is seasonal. Soils on micro-highs next to the pond have thick calcic horizons, and lack argillic horizons (Fig. 14.5; Knuteson et al., 1989).

Site hydrology controls the distribution and genesis of these different soil horizons. Argillic horizons occur beneath recharge ponds. The fresh pond water leaches both ions and clay as it moves to the water table (Knuteson et al., 1989, Table 2). The topographically higher soils next to the recharge ponds have calcic horizons and seldom are water saturated. They receive evaporative discharge (upward unsaturated flow) for several weeks each year from the saturated zone associated with the recharge pond (Knuteson et al., 1989). Calculations of the upward flux of water and its carbonate content show that these calcic horizons (Fig. 14.5) can form within 4,000 to 9,000 years (Knuteson et al., 1989).

Soil properties change with topographic position within the Prairie Pot Hole Region, but similar landforms do not always have a specific soil. Topography only partly controls the direction of local groundwater flow. Groundwater pathways determine many soil physical and chemical properties (Arndt and Richardson, 1988, 1989; Knuteson et al., 1989; Miller et al., 1985; Skarie et al., 1987; Steinwand & Richardson, 1989). Vegetation patterns and growth reflect groundwater movement through the system (Bigler & Richardson, 1984).

A similar but less dramatic relationship between soils and hydrology exists in the southeastern Coastal Plain (Daniels & Gamble, 1967; Daniels, Gamble, Boul et al., 1975; Daniels, Gamble, & Holzhey, 1975; Daniels et al., 1966, 1967, 1977). Position on the landscape and how water moves upon or beneath that landscape controls distribution and properties of fragipan, E, Bt, Bh and organic horizons.

On a nearly level coastal plain landscape away from the dissected edge, the better drained soils are on the micro-highs. Water runs off the high areas during intense storms, and the water table is slightly deeper than under depressions. It is not unusual to have the capillary fringe at or near the surface of the high areas near the watershed divide (Daniels et al., 1987).

Figure 15.13 is a map showing the areal extent of the undissected Coastal Plain surface and soils near Newton Grove, North Carolina. The stream system truncates the surface sediments and exposes a clayey marine unit. Figure 15.12 is the flownet for the area approximately along line A–A′. Yellowish red Bt horizons occur only on the edges of the dissected surface. The distribution of these soils coincides with the deepest water table (Fig. 15.12). The progressive change from yellowish red to strong brown to yellowish brown is common. The clay content of the Bt horizon decreases from the yellowish red to the yellowish brown soils (Fig. 15.14). Iron contents may not vary greatly across the sequence (Daniels & Gamble, 1967, Table 1).

Figure 15.15 is a cross-section of the changes in soils from a micro-high to a lower wetter area near the divide center. The relief is about 1 m in 300 m. The water table is above the surface of the low area for extended periods under natural conditions. The nearby highs seldom have free water closer than 30 cm beneath the surface (Daniels et al., 1987). All soil differences are pedogenic because there is no evidence of a sediment change across the sequence.

The equipotential lines in this area are vertical (Fig. 15.12), so flow should be lateral. A very local flow system may develop between the high and the low because a small head develops during a high stand of groundwater. During the growing season, the water table may drop 1 to 3 m (Daniels et al., 1987), but it is above 50 cm in the low areas for 4 to 6 months each year.

Aeric Ochraquults with sandy loam to loamy sand subhorizons occupy the transition from the topographic highs to the lows. Some of these soils have coarse sandy loam fragipan horizons (Fig. 15.15). The coarse-textured soils in the sequence have high water tables at 50 cm or less from January through May (Daniels et al., 1987, p. 79–86).

The high water table during the growing season suggests that an evaporative upward flux exists for considerable periods. The transition soils have the

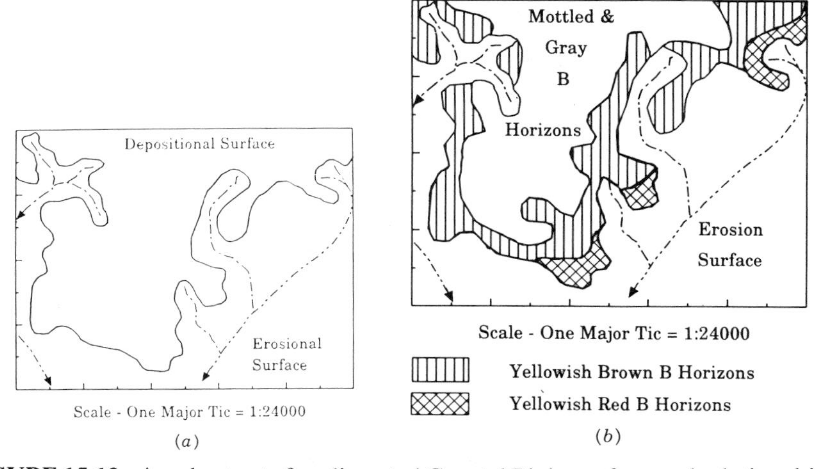

FIGURE 15.13. Areal extent of undissected Coastal Plain surface and relationship of B horizon color to position on the surface. A. Undissected surface. B. Soil distribution.

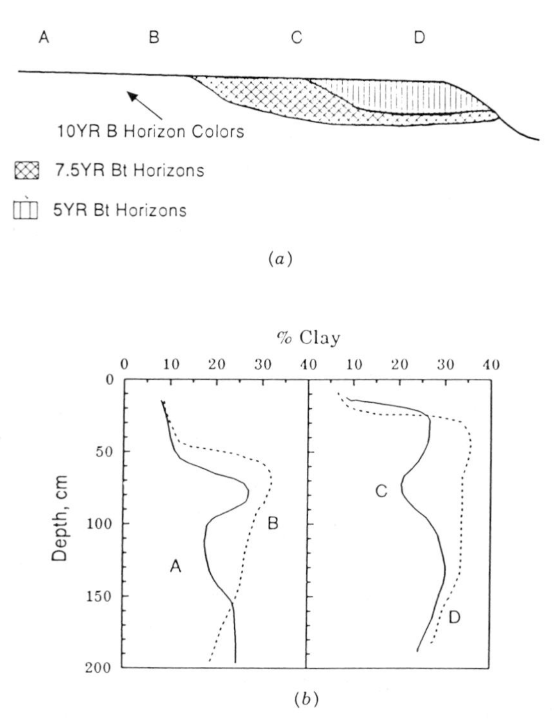

FIGURE 15.14. *Changes in Bt horizon clay content with distance from the dissected edge of a Coastal Plain surface.* Redrawn from Daniels and Gamble, 1967, Newton Grove area of Fig. 3, p. 121. Reprinted by permission from Elsevier Science Publishers.

largest water-table fluctuation. If energy were available, these soils could have several oxidation-reduction cycles each year.

Alternate oxidation and reduction with subsequent clay destruction (Brinkman, 1969–1970, 1973) could explain the change in texture across this soil sequence. Downward movement of the groundwater during the dryer seasons removes the products of clay destruction. Some concentration of material also can occur as the soil dries. Fewer oxidation-reduction cycles in the wetter areas may explain why coarse textures only occur in the transition zone.

The soil sequence in Figure 15.15 reflects the local hydrology and the associated oxidation-reduction regimes. If groundwater flux was downward, a much different suite of soils would exist. We suggest that most soils in the area would have Bt horizons. The wetter sites should have strongly expressed Bt horizons similar to the recharge ponds in North Dakota or the red edge in North Carolina. Similar relationships exist between soils with Bh horizons and their landscape and hydrology (Daniels, Gamble, & Holzhey, 1975; Daniels et al., 1987).

The high-sodium soils of the Macon Ridge (Martin et al., 1980; Weems et al., 1977) are an example of the relationship of soils to regional groundwater hydrology. Fleming (1984) and Goh (1984) studied the dissolution of Na-bearing minerals in the loess as the potential source of sodium. This follows the work of Wilding et al. (1963) in Illinois and Pettry et al. (1981) in Mississippi. Their interpretation was that feldspar weathering and lateral flow from the uplands causes sodium to accumulate in the swales. Their suggestion is reasonable if one looks only at the soils and their properties.

The Mississippi River alluvial aquifer underlies the Macon Ridge loess mantle. In most places, the aquifer has less than 50 mg/l chloride (Whitfield, 1975). A zone of salty water 46 miles long and 1 to 5 miles wide with >100 mg/l chloride occurs under part of the ridge (see Whitfield, 1975, plate 8 and p. 12–

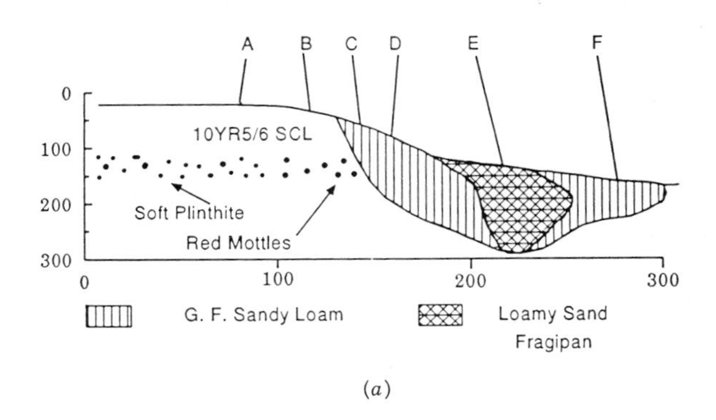

(*a*)

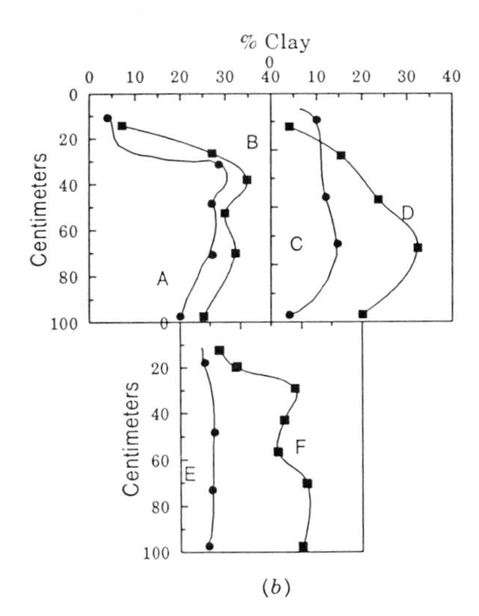

(*b*)

FIGURE 15.15. *Changes in soil morphology and texture across a drainage sequence* at Newton Grove, North Carolina. Redrawn from Daniels and Gamble, 1976, Newton Grove area of Fig. 2, p. 119. Reprinted by permission from Elsevier Science Publishers.

13). The high-sodium soils match the area underlain by salty water (Day & Walthall, 1989; Walthall et al., 1992).

The concentration of Na and Cl from water extracts is similar to those from saline wells. These authors also traced the high levels of Na at depth across both ridge and swale positions. Data of Whitfield (1975), Day and Walthall (1989) and Walthall et al. (1992) are convincing evidence that saline groundwater is the source of Na in these soils.

Similar relationships occur in Saskatchewan, Canada. Henry et al. (1985) showed that discharge of saline groundwater from glacial and bedrock aquifers is responsible for soil salinization. They found that the salt-load could accumulate by upward movement from the aquifer in 500 to 5,300 years.

Soil scientists' training emphasizes the soil profile. This results in interpretation of soil genesis from its physical and chemical properties alone. Any hope of understanding soil development requires consideration of the entire soil environment, including the deeper groundwater. A knowledge of the local hydrogeology is very important in landscapes with saline soils, sediments or aquifers.

HILLSLOPE HYDROLOGY

Rolling or dissected landscapes with deep water tables have a complex, shallow hydrologic system that affects soil morphology and management. The shallow hydrologic system includes temporary saturated zones above the groundwater surface, and downslope transfer of water. Lateral downslope transfer of water (flowthrough) is by a combination of saturated and unsaturated flow. Controlling features are relief, landsurface shape, properties of soil horizons and discontinuities within the upland or hillslope sediments (Anderson & Burt, 1978; Beasley, 1976; Burt & Butcher, 1985; Kirkby, 1988; Lavee et al., 1989; Michiels et al., 1989; O'Loughlin, 1981, 1986; Palkovics & Petersen, 1977; Weyman, 1973).

Change in water content occurs over a few meters horizontally and tens of centimeters vertically. The temporary zones of saturation often are thin, discontinuous and perched above the local, intermediate or regional flow systems (Burt & Butcher, 1985). Only on the foot slopes can these deeper systems merge with the shallow flow system.

Figure 15.16 is a stream hydrograph from a single storm on a small watershed. It shows the delayed storm discharge and the long period of decreasing discharge following the storm. There was no rainfall after the event for the period shown (Anderson & Burt, 1977a, 1977b). Figure 15.17 gives the soil-water tensions (dashed lines) before, during and following the storm. The changes in sizes and locations of saturated zones following the storm show how topography and time influence soil moisture in a rolling landscape.

Zaslavsky and Sinai (1981a) noted concentration of moisture in sand dunes of Northern Sinai. Rainfall is 70 mm/yr and infiltration capacity is near 500

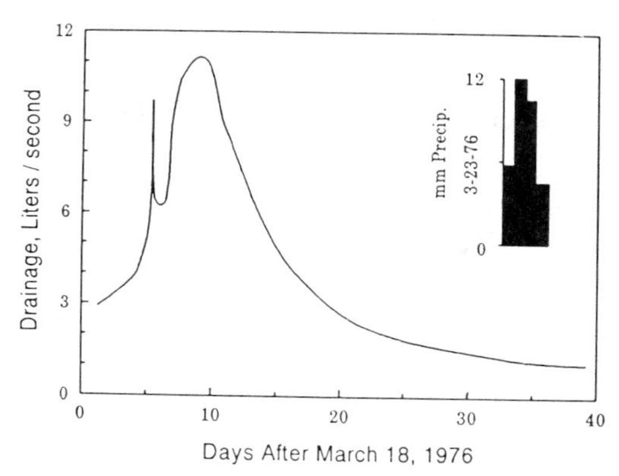

FIGURE 15.16. *Hydrograph of a storm event and resulting discharge.* Redrawn from Anderson and Burt, 1977, Fig. 4, p. 33. Reprinted by permission from Elsevier Scientific Publishing Company and M. Anderson.

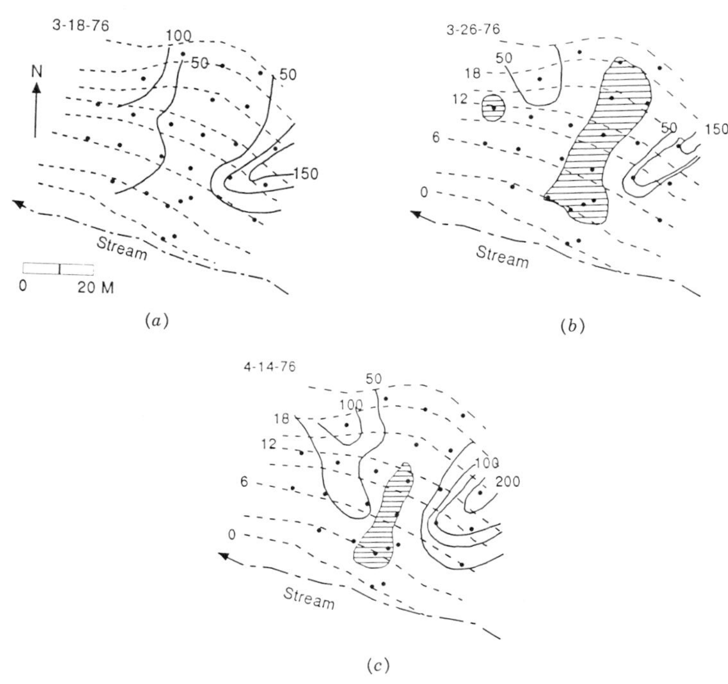

FIGURE 15.17. *Map showing the changes in moisture tension in hollow and spur ridges before, during and after the storm in Figure 15.16.* A: 3–18–76, before rainstorm. B: 3–26–76, during maximum outflow 4 days after rainfall. C: 4–14–76, during outflow recession 13 days after rainfall. Redrawn from Anderson and Burt, 1977, Fig. 5, p. 34. Reprinted by permission form Elsevier Scientific Publishing Company and M. Anderson.

mm/hr. Moisture accumulates in depressions where no surface run-off has occurred and no water table is present. Impermeable layers and perched water tables are absent within these concave sites.

Bedouins use this extra moisture by planting barley only in concave parts of the landscape. Similar landuse occurs in some sand dune areas of Niger where infiltration per hour is greater than or equal to the yearly rainfall. The land operators plant grain sorghum in the swales and depressions and millet on the rest of the landscape.

Most soil landscapes have localized wet areas. Experienced soil surveyors mark these areas with wet-spot or seep symbols. Stratigraphic discontinuities, such as loess over till, may be responsible for some wet areas related to the shallow flow system. The major cause of these localized areas is downslope water transfer produced by differences in permeability between soil horizons. Soil development, roots or soil fauna are among the many factors producing the permeability differences (Beasley, 1976; Zaslavsky & Sinai, 1981a, 1981b).

Theory

Layered soils and sediments control the shape of the stream hydrograph and localized areas of saturation in Figures 15.16 and 15.17. Infiltrating water temporarily perches above the less permeable horizon (Fig. 15.18), and a downslope hydraulic gradient develops. The infiltrating water then moves downslope

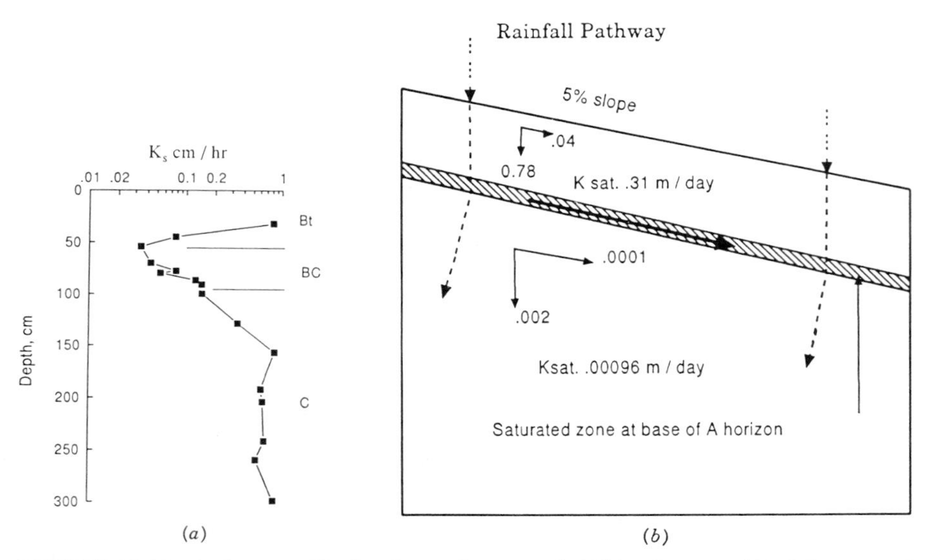

FIGURE 15.18. A. *Saturated hydraulic conductivity of a NC Piedmont soil.* B. *Schematic of rainfall pathways in a sloping Piedmont landscape.* A. Drawn from P. Schonenberger, unpublished data.

above the contact to the restrictive horizon. The subsurface flow mimics the surface contours—it diverges in convex spurs and converges in hollows (head slopes) or concave areas (Fig. 15.17).

Zaslavsky and Sinai (1981a, 1981b) give the reasons for uneven distribution of moisture in a sloping landscape. The saturated and unsaturated hydraulic conductivity of the soil usually changes with depth. If the hydraulic conductivity increases with depth, the streamlines curve downslope to develop a horizontal flow component. If the conductivity decreases with depth, the streamline may turn upslope, but not beyond a direction normal to the soil surface (Zaslavsky & Sinai, 1981b).

The documentation for horizontal flow in sloping landscapes is abundant (Beasley, 1976; Bunting, 1961; Burt & Butcher, 1985; Weyman, 1973). One reason for horizontal flow in many soils is that both saturated and unsaturated hydraulic conductivity decrease with depth because clay content increases downward. Water can accumulate above the horizon with lower hydraulic conductivity during periods of high infiltration. A higher water content increases the horizontal hydraulic conductivity and allows water to move downslope in response to the lateral hydraulic gradient. Water accumulates in the base of the upper horizon and moves downslope in response to the lateral component of the hydraulic gradient.

Figure 15.18 shows the saturated hydraulic conductivity of a soil in the North Carolina Piedmont. The conductivity in the surface horizon is .31 m/d and at 1 m is .00096 m/d. Using these conductivities and a slope of 1 m per 20 m, a length of 20 m, and a porosity of 0.4, we can calculate the vertical and downslope movement of a saturated zone at the base of the A horizon. Equation 3 is from Heath (1987)

$$v = K - n \times (dh) - (dl) \tag{3}$$

where v is velocity; K is hydraulic conductivity; n is porosity; (dh) is vertical elevation change over a given horizontal distance (dl).

Figure 15.18B gives the downslope and vertical velocity of soil-water using equation 3. Although the calculated downslope velocity is not large, it is higher than the vertical velocity in the B–BC horizon. The calculated velocities show the control the restrictive soil horizons have on recharge to the groundwater surface and the potential for downslope movement.

Almost all sloping landscapes have differences in materials related to surface erosion and deposition. Kleiss (1970) named these materials hillslope sediments. Chapter 5 shows progressive changes in thickness and physical properties of these materials with distance from the source. Hydraulic conductivity usually decreases between hillslope sediment and the underlying materials, especially on the upper slopes.

On the upper slopes, the hillslope sediments are coarser than the underlying materials, but median diameter decreases downslope. Hydraulic conductivity should decrease downslope to the concave areas. Sediment thickness above the discontinuity increases downslope, so the volume of soil receiving throughflow also increases.

Horizontal flow starts when water moving downward accumulates above a less permeable horizon. A higher water content increases the horizontal conductivity of that layer and the rate at which water moves. The increased velocity in the more permeable horizon causes the flowlines to bend downslope, and lateral flow develops. Downslope water movement is from saturated or unsaturated flow.

Transfer of water from the higher to lower areas of the landscape results in water accumulating in the lower slopes, especially in the concave areas. Downslope subsurface flow often is at right angles to the surface contours because soil horizon contacts mimic the surface form. For a given width along the contour and an equal distance from the drainage divide, the concave areas have the largest contributing area. The wedge-shaped contributing area of a concave slope is widest upslope (O'Loughlin, 1981). A convex area has a wedge-shaped contributing area with the widest part pointing downslope. Thus one would expect a lower increase in water content downslope in convex than in concave areas.

Saturation can develop in the concave area for several reasons. The most important is the large upslope contributing area that supplies water to an ever-decreasing volume of soil or sediment. The gradient of a concave area decreases downslope, so water outflow is less at the same moisture content and conductivity than sites with more slope (Zaslavsky & Sinai, 1981b). Infiltrating overland flow can also contribute to the saturation of concave areas and lower slopes.

OTHER FACTORS

Unsaturated flow dominates the downslope movement of water (Harr, 1977), although most studies emphasize saturation of parts of the landscape (Anderson & Burt, 1977a,b; Hibbert & Troendle, 1988). Harr measured more downslope flow below 30 cm between storms and more vertical flux at this depth during storms. Unsaturated flow dominated all but the lower 12 to 15 m of the study slope. Helvey et al. (1972) also measured largely unsaturated flow on all but the lower slopes. The volume of water moving horizontally can be large. Gaskin et al. (1989) found that lateral flow averaged 20 to 30 percent of the vertical flow in the A and Ba horizon of a Typic Hapludult.

The idea that unsaturated flow transfers soil-water from the higher to the lower areas of the landscape and can result in local saturation supports field observations. The throughflow mechanism is useful because it helps one predict where saturated soils will occur in the landscape. For example, O'Loughlin (1986) predicted the extent and location of saturated zones on a sloping landscape after timber harvest (Fig. 15.19). Note the changes in both areal extent and moisture content of saturated sectors after harvest in the lower linear and concave slopes.

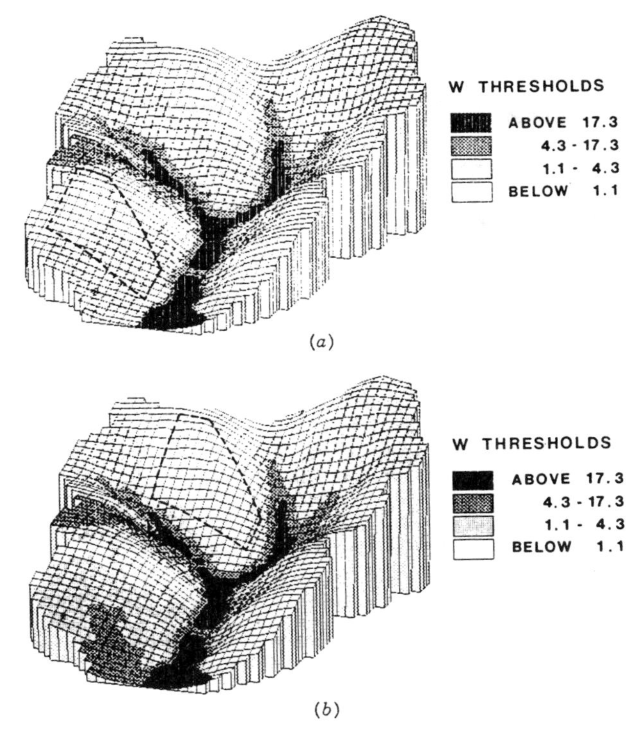

FIGURE 15.19. A. *Predicted zones of surface saturation under full forest cover*. B. *Predicted zones of surface saturation with 90% forest clearing*. The polygon defines a region where drainage flux is reduced one quarter. From O'Loughlin, 1986, A Fig. 4, p. 799, B, Fig. 5a, p. 800. Reprinted by permission from American Geophysical Union.

It is difficult to obtain the research data needed to quantify the changes in moisture shown in Figures 15.16, 15.17 and 15.19. Short-term experiments may measure only small differences. We feel that assuming uniform soils through the study area handicaps these investigations. To collect data from sites that help our understanding of process requires a thorough soils investigation. Topography alone is helpful, but it is only one factor. Careful interpretation of soil morphology and the local stratigraphy will improve the chances of a successful study. One must remember that soil morphology is in part the result of a long-term integration of site hydrology.

APPLICATION TO SOILS

Water drives soil development, plant growth and erosion. Water is the basis for most factors of soil use and formation. We need to think of soil and water as part of a landscape system, not as separate units. Water is the dynamic part. At any one time it may be part of the recharge (leaching), throughflow (transfer) and discharge (addition) hydrologic system in the soil landscape. We cannot study soil development and propose grand schemes of soil genesis without a concurrent study of the soil-landscape hydrologic system.

The idea that saturated and unsaturated flow moves downslope as well as vertically has many implications for most workers in soils. For the individual mapping soils, it helps explain much of the variability in morphology, especially of the A horizon and B horizon mottling patterns. The differences found may be within the range of a soil series, or the soil bodies may be too small to separate at the map scale. Yet the differences are important in soil erosion and productivity studies, and for any land application of waste.

How soil-water moves within and upon the landscape should influence our ideas of soil weathering or development. We train soil scientists to compare the soils in one place with those of another. Field soil scientists in particular constantly need to evaluate and compare soils in landscapes. The tendency to compare the "degree of weathering" of one soil or one horizon with another is both natural and strong, but we muddy our thought processes by using poorly quantified terms such as "more strongly weathered."

How does one separate the effects of intensity of weathering from the effects of duration of weathering? How does one compare the effects of weathering between well-drained and poorly drained soils? Different kinds of weathering exist in different places in soil profiles and in the landscape.

For example, illuvial clays are the product of weathering (dissolution), recrystalization or alteration, transport and deposition. Development of an Alfisol requires several different processes. The accumulation of such products, such as the development of thick argillic horizons downslope, may be as much an indicator of process duration as of intensity. This is particularly true where the products are being concentrated from a larger (upslope) volume of sediments and/or soil.

Comparisons of the degree of weathering among soil horizons or soil profiles should include the following (Birkeland, 1984):

1. The thermodynamics and kinetics of the soil horizons (Scrivner et al., 1973).
2. The frequency and duration of water-movement pulses through the soil profile (Scrivner et al., 1973).
3. The kinds of processes involved (Simonson, 1959).
4. The duration and intensity of the individual processes (Crompton, 1962).
5. The physical and chemical attributes of the parent materials.

We believe that we should limit the use of generalized terms to compare the "intensity" or "degree" of weathering of soil profiles in landscapes. We can justify these generalizations only when available data support the observations. The tendency to generalize may result in misplaced emphasis or erroneous conclusions. One should view the different expressions of development within and among pedons as indicators that soil-forming processes are not uniform within the landscape.

The concept of "weathering" as addressed in the pedological literature is not precise. Applying the terms "more weathered" and "better developed" do not help us understand the measured differences in soil properties that we cannot explain within our knowledge of specific developmental processes. Too often we encounter the idea that warmer climates and longer growing seasons result in "more strongly weathered" or "better developed" soils. These descriptive phrases limit our ability to transfer knowledge, are often misleading, and bias our approach to studying soil-forming processes. These generalizations are very dangerous when applied to soil orders.

For example, let us examine a common climo-toposequence of soils on the western slopes of the Cascade Mountains in Skagit County, Washington. The sequence includes the dryer and warmer Squires map units at lower elevations, and the colder, more moist Clendenen soils at higher elevations (Table 15.1). These soils form in mixtures of loess, volcanic ash, and colluvium over phyllite bedrock. The Clendenen subsoil contains glacial till derived from phyllite. The soil temperature regime is from mesic through cryic as one proceeds up the mountainsides. Annual precipitation is from about 55 inches (Squires) to 90 inches (Clendenen) as the growing season decreases from about 180 to 80 days (Table 15.1).

Clendenen pedons have a visually striking morphology. The thick light-colored E horizon overlies a dark, acid spodic horizon. The Clendenen

TABLE 15.1. Properties and Taxonomic Classifications of a Climo-toposequence of Soils in the Western Cascade Mountains of Washington

Soil Series	Classification	Elevation Range (ft)	Precip. (In.)	Growing Season (Days)	Horizon Sequence
Squires	Medial-skeletal mesic Andic Xerochrepts	400–1,500	55	160–200	Oi, Oe, E, Bw1, Bw2, C1, C2, R
Rinker	Loamy-skeletal frigid, Andic Xerochrepts	1,000–2,200	75	120–150	Oi, Oa, E, Bw1, Bw2, 2C, 2R
Springsteen	Medial-skeletal Typic Cryorthods	2,000–3,100	80	90–120	Oi, Oa, E, Bs1, Bs2, BC, 2C, R
Crinker	Medial-skeletal Typic Cryorthods	2,800–4,000	85	90–110	Oi, Oe, E, Bs1, Bs2, BC, C, R
Clendenen	Medial-skeletal shallow Humic Cryorthods	2,600–4,000	90	80–120	Oi, Oa, E, Bhs, Bs, Bs, 2Cr

From Klungland and McArthur (1989).

pedons have morphologically distinct horizons. Squires pedons have a cambic B horizon, or moderate to poorly differentiated horizons. The cambic B horizon is an indicator of minimal pedogenic development (Soil Survey Staff, 1975). From morphology alone, the Clendenen pedon is "more developed" than the warmer, and possibly older, Squires.

The Squires-Clendenen sequence suggests that soil profile development increases as temperature and growing season decrease. This idea is a challenge to current pedological concepts that use relative degrees of horizon development or soil classification as an indicator of degrees of soil development or process. Many factors are responsible for the morphologic expression of the soil sequence in Table 15.1. For those who must generalize, we suggest Crompton's (1962) essay on soil formation.

Sinai et al. (1981) suggest that the B horizons should be thicker and more strongly weathered where moisture accumulates. Few studies test this idea in sufficient detail within one geomorphic surface on a local landscape. The Burt et al. (1984) data on solutional denudation by landscape position does support the idea of more "weathering" in the moist sites. A continuing problem with studies of this type is separating the effects of differences in materials, local stability and moisture on the properties of the soil profile.

For example, a study by Butler et al. (1986) in the North Dakota Badlands found progressive increases in clay content downslope with a decreasing moisture deficit. This study seems to fit the idea of more strongly expressed Bt horizons with greater moisture supply. The clay increase may be from deposition of finer material as hillslope sediment, or from the outcrop of more clayey parent materials. Some clay increase can be from more favorable moisture relations downslope. Available data do not negate these possibilities. In any study, it will be difficult to separate the various responsible factors.

The transfer of moisture downslope may explain the thin sola of many soils on sharply convex slopes (Bunting, 1961). We usually conclude that erosion is the reason for these thin soils, but low moisture compared to the surrounding area is a factor. The data of Anderson and Burt (1977a) show these slopes are the driest part of the landscape. Soils on convex slopes should have the thinnest A horizons, with lower accumulations of organic matter and related nutrients for the local landscape. Several published articles partially verify the idea (Aandahl, 1948; Aguilar and Heil, 1988a, 1988b; Butler et al., 1986; McCracken et al., 1989).

Soil productivity, and especially the effect of soil erosion, is an important issue, but it is one we feel is often misinterpreted. The differences in moisture content within a local landscape (Figs. 15.17 and 15.19) can lead to large differences in yield (Simmons et al., 1989). Yield differences will vary with crop and year. Abundant or excess moisture during late autumn and winter appears to depress winter grain yields (Miller et al., 1988; Whitman et al., 1985). The effect of higher moisture on corn yields will depend upon the moisture content during the critical periods of tasseling and pollination. Thus, obtaining the necessary data to quantify the relationships to yield will be difficult and should be long-term studies. The dilemma of trying to evaluate the influences of the various factors of soil and micro-climate upon the crop used will always exist.

SUMMARY

How water moves upon and within the landscape has a direct bearing upon most soil properties and their production ability for any given crop. We need more detailed studies of the temporal changes in moisture and its link to soil productivity. The dynamic nature of the processes, and often their short duration, increases the difficulty of obtaining necessary data. Only when we study the soil within its stratigraphic, geomorphic and hydrologic setting can we make major advances in understanding our soils and their potential.

REFERENCES

Aandahl, A.R. (1948). *Soil Sci. Soc. Amer. Proc.*, 13:449–454.

Aguilar, R. and R.D. Heil. (1988a). *Soil Sci. Soc. Am. J.*, 52:1076–1081.

Aguilar, R. and R.D. Heil. (1988b). *Soil Sci. Soc. Am. J.*, 52:1081–1085.

Anderson, M.G. and T.D. Burt. (1977a). *J. Hydrology*, 33:27–36.

Anderson, M.G. and T.D. Burt. (1977b). *J. Hydrology*, 33:383–390.

Anderson, M.G. and T.D. Burt. (1978). *Earth Surface Processes*, 3:331–344.

Arndt, J.L. and J.L. Richardson. (1988). *Wetlands*, 8:93–108.

Arndt, J.L. and J.L. Richardson. (1989). *Soil Sci. Soc. Amer. J.*, 53:848–855.

Beasley, R.S. (1976). *Soil Sci. Soc. Amer. J.*, 40:955–957.

Bigler, R.J. and J.L. Richardson. (1984). *Soil Surv. Horiz.*, 25:16–24.

Birkeland, P.W. (1984). *Soil and Geomorphology*. New York: Oxford Univ. Press.

Brinkman, R. (1969–1970). *Geoderma*, 3:199–206.

Brinkman, R. (1973). *Geoderma*, 10:259–270,.

Bunting, B.T. (1961). *Amer. J. Sci.*, 259:503–518.

Burt, T.P. and D.P. Butcher. (1985). *J. Soil Sci.*, 36:469–486.

Burt, T.P., R.W. Crabtree, and N.A. Fielder. (1984). Patterns of hillslope solutional denudation in relation to the spatial distribution of soil moisture and soil chemistry over a hillslope hollow and spur. In *Catchment Experiments in Fluvial Geomorphology* (pp. 431–445). Ed. by T.P. Burt and D.E. Walling. Norwich: Geo Books.

Butler, J., H. Goetz, and J.L. Richardson. (1986). *Amer. Midland Naturalist*, 116:378–386.

Crompton, E. (1962). *Outlook on Agriculture*, 3:209–218.

Dalrymple, J.B., R.S. Blong, and A. Conacher. (1968). *Annals Geomorphol.*, 12:60–76.

Daniels, R.B. and E.E. Gamble. (1967). *Geoderma*, 1:117–124.

Daniels, R.B., E.E. Gamble, S.W. Buol, and H.H. Bailey. (1975). *Soil Sci. Soc. Amer. Proc.*, 39:335–340.

Daniels, R.B., E.E. Gamble, and C.S. Holzhey. (1975). *Soil Sci. Soc. Amer. Proc.*, 39:1177–1181.

Daniels, R.B., E.E. Gamble, and L.A. Nelson. (1967). *Soil Sci.*, 104:364–369.

Daniels, R.B., E.E. Gamble, and L.A. Nelson. (1971). *Soil Sci. Soc. Amer. Proc.*, 35:781–784.

Daniels, R.B., E.E. Gamble, L.A. Nelson, and A. Weaver. (1987). *Water-Table Levels in Some North Carolina Soils*. USDA Soil Conservation Service. Soil Survey Invest. Rpt. 40.

Daniels, R.B., W.D. Nettleton, R.J. McCracken, and E.E. Gamble. (1966). *Soil Sci. Soc. Amer. Proc.*, 30:376–380.

Daniels, R.B., E.E. Gamble, W.H. Wheeler, and C.S. Holzhey. (1977). *Soil Sci. Soc. Amer. J.*, 41:1175–1180.

Day, W.J. and P.M. Walthall. (1989). *ASA Agron. Abstracts*: 262.

Flemming, B.J. (1984). *A Study of Some Soils Containing High Levels of Exchangeable Sodium in Franklin Parish, Louisiana*. Unpublished master's thesis. Baton Rouge: Louisiana State Univ..

Gaskin, J.W., J.F. Dowd, W.L. Nutter, and W.T. Swank. (1989). *J. Environ. Qual.*, 18:403–410.

Goh, Y.S. (1984). *Natraqualfs and Associated Soils Characteristics and Their Interactions with Gypsum*. Unpublished doctoral dissertation. Baton Rouge: Louisiana State Univ.

Harr, R.D. (1977). *J. Hydrology*, 33:37–58.

Heath, R.C. (1987). *Basic Ground-Water Hydrology*. U.S. Geol. Survey Water-Supply Paper 2220.

Helvey, J.D., J.D. Hewlett, and J.E. Douglass. (1972). *Soil Sci. Soc. Amer. Proc.*, 36:954–959.

Henry, J.L., P.R. Bullock, T.J. Hogg, and L.D. Luba. (1985). *Can. J. Soil Sci.*, 65:749–768.

Hibbert, A.R. and C.A. Troendle. (1988). Streamflow generation by variable source area. In *Forest Hydrology and Ecology at Coweeta* (pp. 111–127). Ed. by W.T. Swank and D.A. Crossley, Jr. New York: Springer-Verlag.

Hursh, C.R. and P.W. Fletcher. (1942). *Soil Sci. Soc. Amer. Proc.*, 7:480–486.

Keller, C.K., G. Van Der Kamp, and J.A. Cherry. (1988). *J. Hydrology*, 101:97–121.

Kirkby, M. (1988). *J. Hydrology*, 100:315–339.

Kleiss, H.J. (1970). *Soil Sci. Soc. Amer. Proc.*, 34:287–290.

Klungland, M. and M. McArthur. (1989). *Soil Survey of Skagit County Area, Washington*. USDA Soil Conservation Service.

Knuteson, J.A., J.L. Richardson, D.D. Patterson, and L. Prunty. (1989). *Soil Sci. Soc. Amer. J.*, 53:495–499.

Lavee, H., M. Wieder, and S. Pariente. (1989). *Earth Surface Process and Landforms*, 14:545–555.

Lissey, A. (1971). *Geol. Assoc. Can. Spec. Pap.*, 9:333–341.

Martin, C.E., L.J. Trahan, and C.T. Midkiff. (1980). *Soil Survey of Franklin Parish, Louisiana*. USDA Soil Conservation Service.

McCracken, R.J., R.B. Daniels, and W.E. Fulcher. (1989). *Soil Sci. Soc. Amer. J.*, 53:1146–1152.

Michiels, P., R. Hartman, and E. De Strooper. (1989). *Earth Surface Processes and Landforms*, 14:533–543.

Miller, J.J., D.F. Action, and R.J. St. Arnaud. (1985). *Can. J. Soil Sci.*, 65:293–307.

Miller, M.P., M.J. Singer, and D.R. Nielsen. (1988). *Soil Sci. Soc. Amer. J.*, 52:1133–1141.

Mills, J.G. and M.A. Zwarich. (1986). *Can. J. Soil Sci.*, 66:121–134.

O'Loughlin, E.M. (1981). *J. Hydrology*, 53:229–246.

O'Loughlin, E.M. (1986). *Water Resources Research*, 22:794–804.

Palkovics, W.E. and G.W. Petersen. (1977). *Soil Sci. Soc. Amer. J.*, 41:394–400.

Pettry, D.E., F.V. Brent, V.E. Nash, and W.M. Koos. (1981). *Soil Sci. Soc. Amer. J.*, 45:578–593.

Richardson, J.L., L.P. Wilding, and R.B. Daniels. (1990). *Recharge and Discharge of Groundwater in the Aquic Moisture Regime Illustrated with Flownet Analysis*. Unpublished manuscript.

Scrivner, C.L.. J.C. Baker, and D.R. Brees. (1973). *Soil Sci.*, 115:213–223.

Simmons, F.W., D.K. Cassel, and R.B. Daniels. (1989). *Soil Sci. Soc. Amer. J.*, 53:534–539.

Simonson, R.W. (1959). *Soil Sci. Soc. Amer. Proc.*, 23:152–156.

Sinai, G., D. Zaslavsky, and P. Golany. (1981). *Soil Sci.*, 132:367–375.

Skarie, R.L., J.L. Richardson, G.J. McCarthy, and A. Maianu. (1987). *Soil Sci. Soc. Amer. J.*, 51:1372–1377.

Steinwand, A.L. and J.L. Richardson. (1989). *Soil Sci. Soc. Amer. J.*, 53:836–842.

Thorpe, J., L.E. Strong, and E.E. Gamble. (1957). *Soil Sci. Soc. Amer. Proc.*, 21:99–102.

Timpson, M.E., J.L. Richardson, L.P. Keller, and G.J. McCarty. (1986). *Soil Sci. Soc. Amer. J.*, 50:490–493.

Walthall, P.M., W.J. Day, and W.J. Autin. (1992). *Soil Sci.*, in press.

Weems, T.A., E.E. Reynolds, E.T. Allen, C.E. Martin, and R.L. Venson. (1977). *Soil Survey of West Carroll Parish, Louisiana*. USDA Soil Conservation Service.

Weyman, D.R. (1973). *J. Hydrology*, 20:267–288.

Whitfield, M.S., Jr. (1975). *Geohydrology and Water Quality of the Mississippi River Alluvial Aquifer, Northeastern Louisiana*. Louisiana Dept. of Public Works. Water Resources Tech. Rpt. 10.

Whitman, C.E., J.L. Hatfield, and R.J. Reginato. (1985). *Agron. J.*, 77:663–669.

Wilding, L.P., R.T. Odell, J.B. Fehrenbaher, and A.H. Beavers. (1963). *Soil Sci. Soc. Amer. Proc.*, 27:432–438.

Zaslavsky, D. and G. Sinai. (1981a). *J. Hydraulics Division, ASCE*, 107:1–16.

Zaslavsky, D. and G. Sinai. (1981b). *J. Hydraulics Division, ASCE*, 107:37–52.

INDEX

Aerosols, 34, 35, 44, 196
Age:
  absolute, 25, 139, 150, 206
  effective, 197, 199
  relative, 14, 15, 16, 17, 18, 25, 139, 150, 196
Aggrading system, 171, 172
Alloclastic, 118
Allophane, 124
Alluvial:
  deposits, 11, 39, 41, 140
  fans, 27, 48, 86
  fill, 6, 18, 79, 164, 165, 171, 173
  flats, 64
  plains, 3, 27, 50, 55, 56
  sediments, 11
  system, 56, 57
Alluviation, 171
Alluvium, 10, 18, 20, 21, 39, 78, 154, 171
Andisol, 115, 124, 125
Aquifer, 213, 217, 218
  confined, 204, 206
  unconfined, 204, 206, 207
Arcuate, 50, 88
Ascendency, 140
Autoclastic, 118
Avalanche scar, 191
Avulsion, 88

Back barrier flat, 103
Backswamp, 48, 55, 56
Backwearing, 148
Barrier, 93, 94, 96, 99
  dunes, 95, 97
  flat, 101, 103
  regressive, 99
  systems, 70, 72, 86, 94, 97, 101, 103, 110, 111
  transgressive, 99
Base level, 146, 147, 149, 151, 154, 155, 156
Beach, face, 107
Beach, ridges, 88, 90, 101, 103, 110
Bedload, 68, 161
Beds, confining, 148, 175, 204, 208
Berm, 107
Biotite, 127, 129, 131
Bioturbated, 109
Bolson, 64, 65
Braided streams, 48, 50, 63
Breccias, 116

Capillary fringe, 214
Channel, 173, 174
  deposits, 50
  ephemeral, 7, 173
Cinders, 115, 119
Clast, 27, 60, 64, 65, 123
Clay:
  illuviation, 201
  maximum, 42
Closed system, 80
Coastal plain, 9, 18, 25, 44, 58, 86, 104, 148, 150, 151, 170
Colluvium, 6, 55, 76, 128, 151, 191
Conductivity, 132, 204, 205, 207, 208, 221, 222
Confining bed, 148, 175
Constructional surface, *see also* Depositional, surface
Convective sinking, 108
Creep, 35, 44, 76, 78, 146, 180, 186, 187, 189, 190, 192
Crevasse splay, 53, 54, 55, 56, 88
Crop yield, 81. *See also* Yield
Cumberland Plateau, 151

Darcey's law, 204
Debris:
  avalanche, 188, 191
  dam, 139, 182
  flows, 60, 63, 64, 188, 189
  slide, 188
Degrading system, 171
Delta, 55, 68, 86, 88, 90, 93, 94, 95, 96, 99, 191
  lake, 65
  plain, 88
Denudation, 146, 147, 159
Depositional:
  environment, 3, 10, 11, 13, 16, 18, 25, 26, 28, 29, 30, 31, 34, 48, 56, 58, 78, 86, 88, 94

Depositional (*Continued*)
  process, 30, 58
  surface, 3, 5, 8, 11, 27, 110, 136, 137, 139, 141, 212
  system, 11, 90, 110, 111
Descendency, 2, 140
Discharge:
  area, 206, 208
  pond, 208, 212, 214
Discontinuity, 77, 78, 147, 222. *See also* Lithologic discontinuity
Dissected surface, 170, 215
Dissection, 9, 148, 149, 152, 157, 169, 170, 175, 177, 201
Dissolution, 134
Dissolved load, 161
Distance from the source, 39, 41
Distributaries, 88
Downwearing, 148
Drainage, 8
Drainage density, 164
Drift:
  glacial, 27, 34
  longshore, 88, 95
  nonstratified, 27
  stratified, 27
Dunes, 34, 35, 36, 43, 58, 64, 65, 68, 95, 97, 107, 218
  barrier, 95, 97
  beach, 95, 110
  coppice, 34, 65
  sand, 34, 36, 65, 68
Dynamic equilibrium, 145, 147, 148, 151

Effective precipitation, 161
Embayments, interdeltaic, 88
Environmental problems, 2
Environments and landforms, 30
Eolian, 27, 34, 35, 36, 39, 43, 44, 50, 64, 68, 70, 140
  deposits, 34
  materials, 34, 44, 99
  processes, 88
  sand, 16, 35, 36, 39, 50, 68, 70
Epiclastic, 118, 119
Equipotential lines, 206, 207, 208, 212, 213, 215
Erosion, 76, 179, 184
  potential, 183
  rates, 134, 146, 159, 160, 161, 162, 166, 167
  surface, 4, 5, 6, 7, 15, 18, 136, 137, 148, 154, 159, 167, 169, 171
Erosional:
  landscapes, 6, 7, 76, 78, 79, 86, 110, 141, 149, 173, 176
  processes, 31, 78, 180
  surfaces, 3, 4, 5, 8, 15, 18, 21, 137, 138, 139, 140, 141, 157, 167
Eruption cloud, 124
Estuaries, 86, 90, 93, 101, 107, 109, 111
Evaporite, minerals, 64, 68

Facies, 18, 50, 86, 99, 108, 110, 111
  beach, 110
  bioturbated mud, 109
  descriptions, 12, 25
  estuarine, 90
  mud, 108, 109
  near shore, 110
  sand ribbon, 108
  sand wave, 108
  shelf, 108, 109
  shoreface, 110
  shore-zone, 94
  storm shelf, 108
Fans, 48, 50, 55, 58, 59, 60, 63, 64, 67, 86, 94, 95, 97, 191
  alluvial, 27, 48
  subaqueous, 86
  tributary, 50
  washover, 94, 97, 99
Faulting, 17, 18, 139
Feldspar, 41, 58, 127, 129, 131, 217
Felsic, 25
Fining upward, 3, 11, 58, 64, 109
Flood channels, 55
Flood plain, 11, 18, 27, 50, 55, 57, 149, 156
Flow, *see also* Mass, movement
  deposits, 63, 64
  earth, 187, 188
  horizontal, 212, 222
  lines (water), 182, 206, 207
  local, 207, 215, 226
  net, 206, 212, 213, 215
  overland, 222
  regional, 207, 218
  saturated, 222, 224
  unsaturated, 214, 218, 222, 224
Fluvial:
  deposits, 35
  environment, 18, 78
  materials, 48
  plains, 3, 56, 57
  processes, 65
  sediments, 3, 11, 50, 70, 107, 140
  system, 48, 57
Foot slope, 5, 6, 48, 84, 182
Freeze–thaw, 185
Frost heave, 78, 186

Geochemical dating, 21
Geologic maps, 3, 11, 12, 34
Geologist, 10, 12, 14, 17, 25, 48, 76, 90, 127
Geology, 10, 12, 13
Geomorphic, 1
  history, 2, 7, 8, 139, 157
  surface, 10, 12, 42, 136, 139, 146, 157, 174, 195, 196, 197, 226
Geomorphology, 1, 2, 3, 7, 8, 10, 12, 42, 136, 144, 156, 203
Geostatistics, 11, 12
Glacial drift, 27
Glacial till, 10, 15, 20, 27–31, 48, 70, 77–80, 157, 165, 172, 212, 214, 220, 225
Glaciers, 27, 48
Gley, 169, 175
Gneiss, 129
Granite, 129
Granitoid source, 63
Ground water, 127, 133, 204, 206, 207, 212–216, 218
  recharge, 207
  saline, 218

Halloysite, 124, 131
Head cut, 79, 155, 170
Head slope, 79
Heave, *see also* Frost heave; Mass, movement
Hillslope:
  evolution, 180, 185
  hydrology, 218
  sediment, 6, 12, 13, 76, 78–81, 84, 89, 149, 155, 218, 221, 226
Holocene, 18, 31, 42, 88
Hornblende, 131
Humate, 101
Hydraulic:
  conductivity, 204, 205, 206, 208, 221
  gradient, 204, 220, 221
  head, 8
Hydroclastic materials, 118
Hydrologic, 9
  cycle, 203
  system, 218, 224
    discharge, 206, 207, 208, 212, 214, 218, 224
    recharge, 206, 207, 208, 212, 214, 216, 221, 223
Hydrology, 1, 2, 3, 5, 7–9, 81, 136, 156, 200, 203, 204, 213, 214, 216–218, 223. *See also* Hillslope, hydrology

Immogolite, 124, 125
Impact energy, 181
Infiltration, 7, 63, 181, 221
Inlets, 94, 95, 96, 97, 99, 111
Intensity of weathering, *see* Weathering, intensity
Interbedded, 41, 64
Interfluve, 5, 79, 128, 141, 154, 161, 164, 166, 170, 171, 173, 197
Interstream divide, 8, 137
  intensity
Iron:
  ferric, 169
  ferrous, 169
  free, 176, 197
Isovolumetric weathering, 161, 190

Kaolinite, 131

Lacustrine, 27, 64, 65, 68–70
  deposits, 68, 69
  sediments, 70
  units, 27
Lag deposits, 11
Lagoon, 86, 93, 94, 95, 96, 101, 107
Lahar, 188
Lakes, 34, 43, 86, 177
  ephemeral, 64, 67, 68, 70
  plain, 70
  saline, 43, 67, 68, 69, 86
Landform, 5, 8, 11, 30, 31, 56, 57, 58, 139
Landscape:
  authors (evolution):
    Dalrymple, 145, 147, 156, 157, 203
    Davis, 145, 146, 147, 150, 157
    Hack, 145–149, 151, 156, 157
    Penck, 145, 146, 147, 150
    Ruhe, 145, 148, 149, 150, 151, 156, 157
  evolution, 144, 145, 159, 161
  position, 2, 81, 86, 156, 157, 226
Landscape surface:
  depositional, 3, 5, 8, 137
  erosional, 3–5, 7–9, 76, 78, 79, 86, 110, 136, 137, 157
Landslides, 159, 189, 191, 192
Lapilli, 119, 124
Lava, 115, 118
Law of ascendency and descendency, 2, 140
Law of superposition, 2, 14, 15
Leaching, 169, 223
Lenticular, 63, 108
Lithologic discontinuity, 77
Lithology, 60
Local thinning, 42
Loess, 3, 4, 10, 11, 16, 20, 21, 27, 34, 35, 39, 41–44, 50, 77–80, 137, 138, 165, 172, 176, 197, 198, 200, 217, 220, 225
Longshore currents, 88, 96, 107

Mafic, 25, 119, 129, 131, 196
Magma, 118, 119
Manganese, 25
Marine:
  deposits, 86, 104
  environment, 26
  shelf sediments, 86, 94, 104, 107, 108, 109, 110
Marsh, 95, 97
  coastal, 94
  salt, 88
Mass:
  movement, 179, 185, 188, 189, 191
  wasting, 185
Meander scars, 50
Mineralogical zones, 45
Moraines, 18, 20, 27, 30
  end, 15
Mud, 109
  bioturbated, 109
  flow, 21, 60, 63, 187, 188, 191
  slump, 187, 191
  spates, 188
Muscovite, 132

Natural levees, 11, 50, 53, 54, 56, 88
Near-shore, 110
Nine-unit landsurface, 145, 149
Nose slopes, 5, 79

Open system, 80
Organic sediment, 69
Organic soils, 9, 70, 104, 174, 177
Outwash, 27, 32, 41, 70
  plain, 20
Overbank deposits, 11, 50
Overland flow, 180, 181–222
Oxidation-reduction, 169, 216

Paleontologist, 12
Paleosol, 39
Parallel slope retreat, 146
Parent materials, 1, 10, 101
Parna, 34, 43, 44
Peat, 88, 148
Pediment, 58, 141, 148, 149
Pedisediment, 141
Pedon, 10, 57, 77, 157
Peneplain, 145, 146, 150
Peneplanation, 145
Permafrost, 187
Permeability, 164
Phi units, 164
Piedmont, 12, 25, 48, 76, 127, 133, 134, 150, 156, 159, 160, 173, 174
Piping, 180
Plain:
  coastal, 3, 18, 25, 44, 86, 104
  flood, 27, 50, 55, 57
Playa, 58, 60, 64, 65, 68
Pocosins, 70
Point bar deposits, 11, 50, 56
Pond:
  flowthrough, 208, 212, 214
  recharge, 212, 214, 216
Predictive model, 185
Preweathered materials, 7
Primary minerals, 35, 58
Principle of flattening, 146
Prodelta, 88
  platform, 90
Pumice, 115, 119, 124
Pseudomorph, 132
Pumice, 115, 119, 124
Pyrite, 129
Pyroclastic deposits:
  fall, 119, 123, 124
  flow, 119, 121, 123, 124
  fragments, 116, 119, 126
  surge, 119, 123, 124

Radiocarbon dating, 21
Radiometric dating, 200
Raindrop, *see also* Impact energy
  impact, 181
  size, 182
Recharge, *see also* Groundwater recharge
  area, 206, 208, 212
  pond, 208, 212, 214, 216, 231
Reduced system, 169
Regolith, 127, 128, 129, 134, 164
Regressive, 15, 99, 109, 111, 116
Relative age, 14, 15, 16, 17, 18, 19, 196
Relict gley, 169
Relict landscape, 146, 148
Relief, 7, 32, 35, 50, 60, 64, 67, 94, 101, 138, 145, 150, 161, 162, 170, 171, 198
Relief/length ratio, 162, 163, 215, 218
Residual soils, 76
Residuum, 128, 133
Ridge:
  sand, 108, 109
River channels, 27
Run-off, 7, 9, 76, 181, 214

Sabkhas, 58, 64, 65
Saltation, 35, 44
Saprolite, 7, 127–129, 131, 133, 134, 144–150, 152–154, 187
Saturated zone, 204, 214, 218, 221, 222

Scarp, 15, 16, 17, 18
Schistose, 129
Scoria, 119
Sediment, 8
colluvial, 76
estuarine, 93
fluvial, 50
glacial, 29, 31
hillslope, 13, 76, 78, 79, 80, 81, 84
inner shelf, 107
load, 50
marine, 25
marsh, 94
organic, 67, 69
shoreface, 107
size, 56
yield, 67, 161, 162, 164, 165
Sedimentation, 25, 50, 58, 67, 68, 99
Shelf:
continental, 104, 107
facies, 108
inner, 107, 108, 109
outer, 108, 109, 110
polar, 107
storm, 108
system, 107
Shoreface, 86, 104, 110, 111
lower, 99, 107, 108
upper, 99, 107
Shore zone, 94
Sieve deposits, 63
Site history, 9, 76, 78, 139, 157
hydrology, 10, 203, 213, 214, 223
position, 6, 9
Slide, *see also* Mass, movement
debris, 188
translational, 188
Slope, 4–7, 21, 39, 42, 43, 48, 55–57, 59, 60, 64, 76, 78–81, 84, 88, 96, 107, 137–139, 146–152, 154–157, 159, 170–174, 176
concave, 184, 222
convex, 184
foot, 5, 6, 48, 84, 182
form, 179
head, 5, 79, 84
linear, 5, 84
nose, 5, 79, 84
shape, 8, 179, 182, 183
shoulder, 5
Slump, 78, 187, 191
Soil:
association, 11, 56
bulk density, 31, 80
development, 1, 9, 201
environment, 199, 218
formation, 1, 3, 7, 36, 43, 169, 195, 196, 199, 200, 201, 226
genesis, 10, 42, 58, 177, 197, 218, 224
horizons, 7, 10, 76
A, 103, 222, 224, 226
argillic, 11, 214, 224
B, 56, 128, 132, 134, 169, 176, 222, 224, 226
Bh, 216
Bt, 11, 131, 133, 200, 215, 216, 226
C, 57, 58, 127, 132
calcic, 214
cambic, 226
E, 225
spodic, 101, 103, 225
landscape, 2, 10, 35, 48, 57, 69, 169, 220, 224
map, 30, 57, 70, 105, 106, 155
mapping unit, 1, 2, 14, 48, 57, 58, 157
materials, 2, 3, 4, 7, 8, 10, 12, 13, 25, 27, 30, 31, 34, 39, 42, 48, 50, 55, 56, 58, 63, 67, 68, 69, 70, 76, 78, 104, 110
morphology, 36, 38, 169, 188, 190, 195, 197, 198, 200, 213, 214, 218
order, 48, 223, 225
organic matter, 67, 70
properties, 1, 2, 7, 8, 11, 25, 31, 46, 76, 86, 139, 150, 157, 195, 213, 214, 225, 227
scientist, 1, 2, 10, 11, 13, 25, 29, 48, 76, 77, 111, 139, 195, 197, 201, 203, 213, 218, 224
sequence, 216, 226
series, 1, 2, 30, 36, 77, 157, 224
solum, 34, 58, 129, 224
system, 8
taxonomy, 10, 25, 57, 76, 77
texture, 7, 11, 25, 31, 35, 56, 103
variability, 10, 11, 12, 25, 42, 48, 57, 58,76
water movement, 12
Solifluction, 187
Solution, 132, 134, 146, 161, 180
Spits, 68
Splays, 50, 53, 54, 55, 56, 88
Spodic, 101, 103, 225
Stone line, 78
Stranding, 63
Straths, 18
Stratigraphic:
column, 3, 13, 123
unit, 3, 13, 42, 78
Stratigraphy, 1, 2, 3, 8, 10, 12, 13, 14, 16, 25, 42, 136, 156, 157, 195, 203, 223
Stream, 3
braided, 48, 50, 63, 65
channel, 48, 170, 172, 173
ephemeral, 64, 65, 173
dissection, 9, 169, 175, 177
hydrograph, 218, 220

Stream (*Continued*)
- incision, 169, 174, 176
- intermittent, 58, 59
- flow, 50, 60, 63
- lines, 221
- meandering, 48, 50, 56, 67
- profile, 173
- systems, 55, 56, 148, 161, 170, 171, 173, 175, 177, 192
- trenching, 171

Subaerial weathering, 6, 200
Subdelta, 88
Subsoil, 128
Superposition, 2, 14, 15, 16
Surface shape, 42
Suspended load, 68
Swash zone, 107

Tephra, 24, 30, 41, 118, 124
Terrace, 17, 20, 25, 50, 57, 148
- fluvial, 15, 17, 18
- marine, 15

Terracette, 187
Terrestrial, 26, 27, 140
Texture, 1, 3, 4, 7, 8, 25, 27, 31, 35, 36, 39
Thalweg, 151, 173
Throughflow, 218, 222, 223
- pond, 208, 212, 214

Tidal:
- deposits, 90
- flats, 94, 95, 109
- range, 95, 96, 97

Tides, 88, 90, 94, 108, 109, 110
Till, 10, 15, 20, 27–31, 48, 70, 77–80, 157, 165, 172, 212, 214, 220, 225
- ablation, 29
- basal, 29, 30
- lodgement, 29
- plains, 15, 78

Time (as a factor of soil formation), 1, 7
- effective, 197
- geologic, 197
- pedogenic, 195
- relative, 18

Toe slope, 6, 78
Topographic position, 128
Topography, 1, 29, 30, 34, 36, 56, 78, 129
Toposequence, 200
Traction deposits, 34
Transect, 8, 57, 197
Transgressive, 15, 97, 99, 107, 108, 109, 111
Transitional environment, 26
Trenching system, 171
Truncation, 11, 16, 18, 42

Ultramafic, 129
Undissected surface, 170
Universal soil loss equation, 154, 183
Unsaturated zone, 204
Upward coarsening, 92, 99, 109, 110

Valley:
- deposits, 141
- floor, 50, 59, 146, 149, 156
- slope, 4, 5, 21, 42, 43, 55, 57, 76, 78, 146–149, 151, 154, 155, 156, 164, 165, 166, 167, 170, 171, 172, 174, 176
- walls, 55

Variability:
- textural, 3, 4, 8, 57

Volcanic:
- ejecta, 115
- materials, 63, 115, 118
- processes, 124

Volcanic deposits:
- fall, 119, 124
- flow, 115, 119, 121, 123
- pyroclastic, 115, 119
- surge, 123

Volcaniclastic, 115, 118

Water, *see also* Ground water; Water table
Water movement, 12, 103
Watershed, 162, 164
Water table, 169, 176, 204, 206, 207, 212, 214, 215, 218, 220
Watershed relief/length ratio, 162
Weatherable minerals, 41, 58
Weathering, 5, 6, 8, 35, 36, 41, 42, 58, 127, 134, 224, 225
- intensity, 197, 199, 201
- isovolumetric, 161, 190
- processes, 224, 225
- profile, 13, 127
- subaerial, 6, 200
- zonation, 137, 139
- zones, 6, 42, 129, 144, 195

Wetland, 212, 214
Winnowing, 35, 107, 139

Yield:
- crop, 81, 227
- grain, 81, 226